MATERIAL BALANCE AND PROCESS CALCULATIONS (REVISED)

A Book for Chemical Engineers and Chemists

By

Kingsley Augustine

TABLE OF CONTENT

- CHAPTER 1 MOLE FRACTION AND MASS FRACTION 3
- CHAPTER 2 AVERAGE MOLECULAR MASS 13
- CHAPTER 3 MATERIAL BALANCE: INTRODUCTION 20
- CHAPTER 4 BALANCES INVOLVING DRYING/EVAPORATIVE PROCESSES 21
- CHAPTER 5 BALANCES INVOLVING MIXING OF SOLUTIONS 30
- CHAPTER 6 BALANCES INVOLVING COMBUSTION 40
- CHAPTER 7 BALANCES INVOLVING LIMITING REACTANTS 59
- CHAPTER 8 BALANCES ON SEPARATION PROCESSES 73
- CHAPTER 9 BALANCES ON SOLVENT EXTRACTION 91
- CHAPTER 10 CALCULATIONS INVOLVING THE DETERMINATION OF FORMULA OF COMPOUNDS 106
- CHAPTER 11 PRESSURE IN LIQUID 120
- CHAPTER 12 HUMIDITY AND WATER VAPOUR IN THE AIR 130
- CHAPTER 13 EQUILIBRIUM REACTION CALCULATIONS 142
- ANSWERS TO THE EXERCISES 172

CHAPTER 1
MOLE FRACTION AND MASS FRACTION

The number of moles of a substance is the ratio of the mass of the substance to the molar mass of the substance. It is expressed as:

$$\text{Number of moles} = \frac{\text{mass of substance}}{\text{molar mass of substance}}$$

Mole fraction is the ratio of the number of moles of any component in a mixture to the total number of moles of all the components in the mixture. It is given by:

$$\text{Mole ratio, } y = \frac{\text{number of moles of any component}}{\text{total number of moles of all components}}$$

Mass fraction is the ratio of the mass of any component in a mixture to the total mass of all the components in the mixture. It is given by:

$$\text{Mass fraction, } x = \frac{\text{mass of any component}}{\text{total mass of all the components}}$$

Note that the sum of the mole fractions of all the components in a mixture is equal to one.

Similarly, the sum of the mass fractions of all the components in a mixture is equal to one

Thus, if y is the mole fraction and x is a mass fraction, then symbolically, we have that:

$$\sum y = 1 \quad \text{and} \quad \sum x = 1$$

where the symbol, $\sum$, means summation.

Examples

1. A mixture contains 4.2 kg of X, 1.8 kg of Y and 3.5 kg of Z. The molecular mass of X, Y and Z are 16kg/kmol, 44kg/kmol and 28kg/kmol respectively. Calculate:

(a). The number of moles of each component

(b). The mole fraction of each component

(c). The mass fraction of each component

Solutions

(a). Number of kmols = $\dfrac{\text{mass of component}}{\text{molar mass of component}}$

Therefore, number of kmols of X = $\dfrac{4.2}{16}$

$\qquad\qquad\qquad\qquad$ = 0.2625 kmols

Number of kmols of Y = $\dfrac{1.8}{44}$

$\qquad\qquad\qquad\qquad$ = 0.0409 kmols

Number of kmols of Z = $\dfrac{3.5}{28}$

$\qquad\qquad\qquad\qquad$ = 0.1250 kmols

(b). Total kmols of all components = 0.2625 + 0.0409 + 0.125

$\qquad\qquad\qquad\qquad\qquad\qquad$ = 0.4284

Mole fraction = $\dfrac{\text{Number of moles of a component}}{\text{total number on moles of all components}}$

Therefore, mole fraction of X = $\dfrac{0.2625}{0.4284}$

$\qquad\qquad\qquad\qquad$ = 0.613

Mole fraction of Y = $\dfrac{0.0409}{0.4284}$

$\qquad\qquad\qquad\qquad$ = 0.095

mole fraction of Z = $\dfrac{0.125}{0.4284}$

$\qquad\qquad\qquad\qquad$ = 0.292

(c). The total mass of all components = 4.2 + 1.8 + 3.5

$\qquad\qquad\qquad\qquad\qquad\qquad$ = 9.5 kg

Therefore mass fraction of X = $\dfrac{4.2}{9.5}$

$\qquad\qquad\qquad\qquad$ = 0.442

Mass fraction of Y = $\dfrac{1.8}{9.5}$

= 0.189

Mass fraction of X = $\dfrac{3.5}{9.5}$

= 0.368

2. A Liquefied petroleum gas (LPG) was analyzed to contain the following composition by volume: 52.4% CH_4, 30.9% C_4H_{10} and 16.7% C_3H_8. Calculate:

(a). the mole fraction of each component

(b). the mass of each component

(c). the mass fraction of each component

Solutions

(a). The composition by volume of the component is also regarded as the number of moles of the components.

Therefore, the LPG contains 52.4 moles of CH_4, 30.9 moles of C_4H_{10} and 16.7 moles of C_3H_8

Total number of moles = 100 moles (i.e. 30.9 + 52.4 + 16.7 = 100)

Therefore mole fraction of CH_4 = $\dfrac{52.4}{100}$

= 0.524

Mole fraction of C_4H_{10} = $\dfrac{30.9}{100}$

= 0.309

Mole fraction of C_3H_8 = $\dfrac{16.7}{100}$

= 0.167

(b). Recall that: Number of moles = $\dfrac{Mass}{Molar\ mass}$

Therefore, mass = molar mass x number of moles   (when we cross multiply)

Therefore mass of CH_4 = (12 + 4) x 52.4   (12 + 4 = 16 = molar mass of CH_4)

$$= 16 \times 52.4$$

$$= 838.4 g$$

Mass of C_4H_{10} = [(12 x 4) + (1 x 10)] x 30.9

$$= 58 \times 30.9 \quad \text{(Molar mass of } C_4H_{10} = 58\text{)}$$

$$= 1792.2 g$$

Mass of C_3H_8 = [(12 x 3) + (1 x 8)] x 16.7

$$= 44 \times 16.7$$

$$= 734.8 g$$

(c). Total mass of the hydrocarbon = 838.4 + 1792.2 + 734.8

$$= 3365.4 g$$

Therefore, mass fraction of $CH_4 = \dfrac{838.4}{3365.4}$

$$= 0.249$$

Mass fraction of $C_4H_{10} = \dfrac{1792.2}{3365.4}$

$$= 0.533$$

Mass fraction of $C_3H_8 = \dfrac{734.8}{3365.4}$

$$= 0.218$$

3. A hydrocarbon was burnt in excess air to produce a mixture of gases with the following composition: 1.24 moles of CO_2, 0.28 mole of CO, 0.91 mole of O_2 and 5.66 of N_2. Determine the composition by mass of each component.

Solution

Mass of component = number of moles x molar mass

Therefore mass of CO_2 = 1.24 x [12 + (6 x 2)]

$$= 1.24 \times 44$$
$$= 54.56g$$

Mass of CO $= 0.28 \times (12 + 16)$
$$= 0.28 \times 28$$
$$= 7.84g$$

Mass of $O_2 = 0.91 \times (16 \times 2)$
$$= 0.91 \times 32$$
$$= 29.12g$$

Mass of $N_2 = 5.66 \times (14 \times 2)$
$$= 5.66 \times 28$$
$$= 158.48g$$

Note that the mass fraction of each component can be calculated as follows:

Total mass of all components = 54.56 + 7.84 + 29.12 + 158.48
$$= 250g$$

Therefore mass fraction of $CO_2 = \dfrac{54.56}{250}$
$$= 0.218$$

Mass fraction of CO $= \dfrac{7.84}{250}$
$$= 0.031$$

Mass fraction of $O_2 = \dfrac{29.12}{250}$
$$= 0.116$$

Mass fraction of $N_2 = \dfrac{158.48}{250}$
$$= 0.634$$

4. On analysis, a gaseous mixture of CO_2 and CO was found to weigh 138g. If the total number of moles of the mixture is 4.4 moles, calculate:

(a). The mass of each of CO_2 and CO in the mixture

(b). The mole fraction of each of the components

Solution

(a). Let the mass of CO_2 in the mixture be m. Therefore the mass of CO will be 138 - m

The molecular mass of CO_2 = 12 + (16 x 2)

$$= 12 + 32$$

$$= 44g/mol \text{ (It can also be expressed as 44kg/kmol)}$$

The molecular mass of CO = 12 + 16 = 28g/mol

Therefore, the number of moles of $CO_2 = \dfrac{m}{44}$ (since mass of CO_2 = m)

The number of moles of CO = $\dfrac{138 - m}{28}$

But the total number of moles is 4.4moles. This means that:

$$\dfrac{m}{44} + \dfrac{138 - m}{28} = 4.4$$

In order to solve this equation, multiply each term by 308, (i.e. the L.C.M of 44 and 28). This gives:

$$308(\dfrac{m}{44}) + 308(\dfrac{138 - m}{28}) = 308 \times 4.4$$

$$7m + 11(138 - m) = 1355.2 \quad \text{(Note that } \dfrac{308}{44} = 7, \text{ and } \dfrac{308}{28} = 11)$$

$$7m + 1518 - 11m = 1355.2$$

$$1518 - 1355.2 = 11m - 7m$$

$$162.8 = 4m$$

Therefore, $\quad m = \dfrac{162.8}{4}$

$$m = 40.7g$$

Hence, the mass of CO_2 in the mixture is 40.7g

And the mass of CO in the mixture is 138 - m = 138 - 40.7

$$= 97.3g$$

(b). Number of moles of CO_2 = 40.7/44 = 0.925moles

$$\text{Number of moles of CO} = \frac{97.3}{28}$$

$$= 3.475 \text{moles}$$

The total number of moles = 4.4moles

$$\text{Therefore mole fraction of } CO_2, y_{CO_2} = \frac{0.925}{4.4}$$

$$= 0.210$$

$$\text{Mole fraction of CO}, y_{CO} = \frac{3.475}{4.4}$$

$$= 0.790$$

5. 45g of a gaseous mixture has the following composition: A = 0.75 moles and a mass fraction of 0.22, B = 0.32 moles and a mass fraction of 0.48, C = 0.94 moles and a mass fraction of 0.3. Calculate the molecular mass of each of A, B and C.

Solutions

$$\text{Mass fraction} = \frac{\text{Mass of component}}{\text{Total mass}}$$

Therefore, for A, $0.22 = \frac{\text{mass of A}}{45}$

Hence, mass of A = 0.22 x 45   (when we cross multiply the above equation)

$$= 9.9g$$

Similarly, mass of B = 0.48 x 45

$$= 21.6g$$

And, mass of C = 0.3 x 45

$$= 13.5g$$

Recall that: Number of moles = $\dfrac{\text{Mass of substance}}{\text{Molar mass of substance}}$

Therefore, for A, $0.75 = \dfrac{9.9}{\text{molecular mass}}$

Hence, molecular mass of A = $\dfrac{9.9}{0.75}$

$$= 13.2 \text{g/mol}$$

Similarly, molecular mass of B = $\dfrac{\text{mass of B}}{\text{number of moles of B}}$

$$= \dfrac{21.6}{0.32}$$

$$= 67.5 \text{g/mol}$$

And, molecular mass of C = $\dfrac{13.5}{0.94}$

$$= 14.4 \text{g/mol}$$

6. A glucose solution contains 32wt % of glucose ($C_6H_{12}O_6$). Calculate the mole fraction of glucose in the solution.

Solution

From the question, we can deduce that for a 100g (i.e. from 100%) of solution, the mass of glucose is 32g, while the mass of water is 100 - 32 = 68g.

Therefore, number of mole of glucose = $\dfrac{\text{mass of glucose}}{\text{molar mass of glucose}}$

$$= \dfrac{32}{[(12 \times 6) + (1 \times 12) + (16 \times 6)]}$$

$$= \dfrac{32}{72 + 12 + 96)}$$

$$= \dfrac{32}{180}$$

$$= 0.1778$$

Similarly, number of moles of water = $\frac{68}{18}$ (Molecular mass of water = 18)

$$= 3.778$$

Therefore, mole fraction of glucose = $\frac{0.1778}{0.1778 + 3.778}$

$$= \frac{0.1778}{3.9558}$$

$$= 0.0449$$

Therefore mole fraction of glucose in the solution is 0.0449.

EXERCISE

1. A mixture contains 5.3 kg of X, 2.5 kg of Y and 4.1 kg of Z. The molecular mass of X, Y and Z are 23kg/kmol, 31kg/kmol and 39kg/kmol respectively. Calculate:

(a). The number of moles of each component

(b). The mole fraction of each component

(c). The mass fraction of each component

2. A fuel was analyzed to contain the following composition by volume: 42.2% CH_4, 40.9% C_4H_{10} and 16.9% C_3H_8. Calculate:

(a). the mole fraction of each component

(b). the mass of each component

(c). the mass fraction of each component

(C = 12, H = 1)

3. A hydrocarbon was burnt in excess air to produce a mixture of gases with the following composition: 2.4 moles of CO_2, 1.5 moles of CO, 3.2 moles of O_2 and 6.8 moles of N_2. Determine the composition by mass of each component.
(C = 12, H = 1, O = 16, N = 14)

4. On analysis, a gaseous mixture of NO_2 and NO was found to weigh 92g. If the total number of moles of the mixture is 2.6 moles, calculate:

(a). The mass of each of NO_2 and NO in the mixture

(b). The mole fraction of each of the component

(N = 14, O = 16)

5. 120g of a gaseous mixture has the following composition: X =1.55 moles and a mass fraction of 0.12, Y = 0.81 moles and a mass fraction of 0.55, Z = 2.32 moles and a mass fraction of 0.33. Calculate the molecular mass of each of X, Y and Z.

6. A solution of tetraoxosulphate (VI) acid contains 68wt % of the acid. Calculate the mole fraction of the acid in the solution.
(H = 1, S= 32, O = 16)

CHAPTER 2
AVERAGE MOLECULAR MASS

The average molecular mass of a mixture is the ratio of the mass of the mixture to the number of moles of all components in the mixture. It is given in terms of the mole fraction and molecular mass of each component as follows:

$M_{av} = y_1M_1 + y_2M_2 + y_3M_3 + \ldots y_tM_t$

where M_{av} = average molecular mass of mixture, y = mole fraction of component and M = molecular mass of component and there are t components in the mixture.

It can be represented symbolically as:

$M_{av} = \sum yM$

The average molecular mass of a mixture can also be given in terms of the mass fraction as follows:

$$\frac{1}{M_{av}} = \frac{x_1}{M_1} + \frac{x_2}{M_2} + \frac{x_3}{M_3} + \ldots \frac{x_t}{M_t}$$

where x = mass fraction of component.

Symbolically it can be expressed as:

$$\frac{1}{M_{av}} = \sum \frac{x}{M}$$

Examples

1. A mixture of gases is composed of 0.25 moles of N_2, 1.32 moles of CO and 0.71 moles of Ne. Calculate the average molecular mass of the mixture.

Solution

Total number of moles of the components = 0.25 + 1.32 + 0.71

= 2.28

Therefore, mole fraction of N_2 = $\frac{0.25}{2.28}$

= 0.110

Mole fraction of CO = $\dfrac{1.32}{2.28}$

= 0.579

Mole fraction of Ne = $\dfrac{0.71}{2.28}$

= 0.311

Therefore, $M_{av} = \sum yM$

$= (yM)_{N_2} + (yM)_{CO} + (yM)_{Ne}$

Note that the molecular mass of N_2 = 14 x 2 = 28

The molecular mass of CO = 12 + 16 = 28

The atomic mass of Ne = 20

Therefore, $M_{av} = (yM)_{N_2} + (yM)_{CO} + (yM)_{Ne}$

= (0.110 x 28) + (0.579 x 28) + (0.311 x 20)

= 3.08 + 16.212 + 6.22

= 25.512

Therefore, the average molecular mass of the mixture is 25.5g/mol

2. The components of a mixture are 1.5g of H_2, 0.95g of CO_2 and 0.22g of N_2. Determine the average molecular mass of the mixture.

<u>Solution</u>

Total mass of the components = 1.5 + 0.95 + 0.22

= 2.67g

Therefore, mass fraction of H_2 = $\dfrac{1.5}{2.67}$

= 0.56

Mass fraction of CO_2 = $\dfrac{0.95}{2.67}$

$$= 0.36$$

Mass fraction of $N_2 = \dfrac{0.22}{2.67}$

$$= 0.08$$

The molecular mass of $H_2 = 1 \times 2 = 2$

The molecular mass of $CO_2 = 12 + (16 \times 2)$

$$= 12 + 32 = 44$$

The molecular mass of $N_2 = 14 \times 2 = 28$

Therefore, $\dfrac{1}{M_{av}} = \sum \dfrac{x}{M}$

$$= \left(\dfrac{x}{M}\right)_{H_2} + \left(\dfrac{x}{M}\right)_{CO_2} + \left(\dfrac{x}{M}\right)_{N_2}$$

$$= \dfrac{0.56}{2} + \dfrac{0.36}{44} + \dfrac{0.08}{28}$$

$$= 0.28 + 0.0082 + 0.0029$$

$$\dfrac{1}{M_{av}} = 0.2911$$

Therefore, $M_{av} = \dfrac{1}{0.2911}$

$$M_{av} = 3.44$$

Therefore, the average molecular mass of the mixture is 3.44g/mol

3. A gaseous mixture contains O_2 and an unknown gas. If the average molecular mass of the mixture is 23g/mol and the mole fraction of O_2 in the mixture is 0.45, calculate the molecular mass of the unknown gas.

Solution

Recall that sum of mole fraction of all components of a mixture is = 1

Therefore, $y_{O_2} + y_A = 1$ (where A is the unknown gas)

$$0.45 + y_A = 1$$

$$y_A = 1 - 0.45$$

$$y_A = 0.55$$

Molecular mass of O_2 = 16 x 2 = 32

Therefore, $M_{av} = \Sigma yM$

$$M_{av} = (yM)_{O_2} + (yM)_A$$

$$23 = (0.45 \times 32) + (0.55 M_A) \quad (M_A \text{ is the molecular mass of A})$$

$$23 = 14.4 + 0.55 M_A$$

$$23 - 14.4 = 0.55 M_A$$

$$8.6 = 0.55 M_A$$

Therefore, $M_A = \dfrac{8.6}{0.55}$

$$= 15.6$$

Therefore, the molecular mass of the unknown gas is 15.6g/mol

4. A gas is composed of N_2 and Ne. If the average molecular mass of the gas is 21.5g/mol, calculate the mass fraction of N_2 and Ne in the mixture.

Solution

Molecular mass of N_2 = 14 x 2 = 28

Molecular mass of Ne = 20

Let the mass fraction of N_2 be x.

Therefore the mass fraction of Ne = 1 - x   (since sum of mass fractions is 1)

In terms of mass fraction:

$$\dfrac{1}{M_{av}} = \Sigma \dfrac{x}{M}$$

$$\frac{1}{M_{av}} = \left(\frac{x}{M}\right)_{N_2} + \left(\frac{x}{M}\right)_{Ne}$$

$$\frac{1}{21.5} = \frac{x}{28} + \frac{1-x}{20}$$

$$0.0465 = \frac{x}{28} + \frac{1-x}{20}$$

In order to clear the fractions, multiply each term by 140 (i.e. the LCM of 28 and 20)

$$140 \times 0.0465 = 140\left(\frac{x}{28}\right) + 140\left(\frac{1-x}{20}\right)$$

$6.51 = 5x + 7(1 - x)$ (Note that $\frac{140}{28} = 5$, and $\frac{140}{20} = 7$)

$6.51 = 5x + 7 - 7x$

$7x - 5x = 7 - 6.51$

$2x = 0.49$

Therefore, $x = \frac{0.49}{2}$

$x = 0.245$

Therefore the mass fraction of N_2 = 0.245, while the mass fraction of Ne = 1 - x = 1 - 0.245 = 0.755.

5. The Composition of air by mass is 23.2% of O_2 and 76.8% of N_2, while the composition of air by volume is 21% of O_2 and 79% of N_2. Calculate the average molecular mass of air.

Solution

Method 1: Use of mass fraction.

Total mass percent of the components = 23.2 + 76.8 = 100

Therefore, mass fraction of O_2 = $\frac{23.2}{100}$

= 0.232

Mass fraction of N_2 = $\frac{76.8}{100}$

= 0.768

Therefore, $\dfrac{1}{M_{av}} = \left(\dfrac{x}{M}\right)_{O_2} + \left(\dfrac{x}{M}\right)_{N_2}$

$= \dfrac{0.232}{32} + \dfrac{0.768}{28}$ (Molecular mass of O_2 = 32, while that of N_2 = 28)

= 0.00725 + 0.02743

$\dfrac{1}{M_{av}} = 0.03468$

Therefore, $M_{av} = \dfrac{1}{0.03468}$

M_{av} = 28.8g/mol

Method 2: Use of mole fraction.

Total mole (volume) percent of mixture components = 21 + 79 = 100

Therefore, mole fraction of $O_2 = \dfrac{21}{100}$

= 0.21

Mole fraction of $N_2 = \dfrac{79}{100}$

= 0.79

Therefore, $M_{av} = (yM)_{O_2} + (yM)_{N_2}$

= (0.21 x 32) + (0.79 x 28)

= 6.72 + 22.12

M_{av} = 28.8

Therefore the average molecular mass of air is 28.8g/mol.

EXERCISE

1. A mixture of gases is composed of 0.58 moles of CO_2, 1.95 moles of NO and 3.25 moles of O_2. Calculate the average molecular mass of the mixture.
(C = 12, O = 16, N = 14)

2. The components of a mixture are 10g of H_2, 25g of NO_2 and 7.8g of N_2. Determine the average molecular mass of the mixture.
(H = 1, N = 14, O = 16)

3. A gaseous mixture contains N_2 and an unknown gas. If the average molecular mass of the mixture is 52g/mol and the mole fraction of N_2 in the mixture is 0.85, calculate the molecular mass of the unknown gas.
(N = 14)

4. A gas is composed of Cl_2 and Ar. If the average molecular mass of the gas is 50g/mol, calculate the mass fraction of Cl_2 and Ar in the mixture.
(Cl = 35.5, Ar = 40)

5. The Composition of air by mass is 23% of O_2 and 77% of N_2, while the composition of air by volume is 20.8% of O_2 and 79.2% of N_2. Calculate the average molecular mass of air.
(O = 16, N = 14)

CHAPTER 3
MATERIAL BALANCE: INTRODUCTION

A material balance is an expression indicating the law of conservation of matter which account for all the materials in a process.

It can be expressed as:

    Input - Output = Accumulation

In a process where a chemical reaction occurs, the material balance equation becomes:

    Input - Output = Generation by reaction + Accumulation

When the process is a steady state process where there is no accumulation, then the material balance expression is given by:

    Input - Output = Generation by reaction

Note that when a component goes into a process and does not undergo any change, then:

    Input = Output

This is a case where there is no accumulation and no reaction.

Calculations in material balances are usually carried out in masses, which also involves mass flow rate. The use of mass fraction is also very important. Volumetric flow rate and mole fraction are also useful for material balance calculations.

CHAPTER 4
BALANCES INVOLVING DRYING/EVAPORATIVE PROCESSES

In processes involving drying or evaporation, one of the components usually passes the process without undergoing any change. Only water is removed from the wet solid.

Examples

1. 800kg of wet flour is fed into a dryer. The wet flour contains 40% by mass of water, while the dry product is to contain 5% by mass of water. Calculate the mass of water removed by the dryer.

Solution

Let F = mass of wet flour (i.e. feed)

W = mass of water removed

and P = mass of product (dried flour)

Also, let x_F = % of water wet flour (mass fraction of water in feed)

x_W = % of water in water (this must be 1)

and x_P = % of water in product

Therefore, the overall component balance is given by:

Input = Output   (Since there is no accumulation)

F = W + P ....Equation 1

The water component balance is given by:

Water in wet flour = water removed + water in product

Therefore, $Fx_F = Wx_W + Px_P$ ....Equation 2

Note that amount of water in a substance = % of water in the substance x mass of the substance

Therefore from equation 1,

$$F = W + P$$

$$800 = W + P \quad \text{...............Equation 3}$$

From equation 2:

$$Fx_F = Wx_W + Px_P$$

(Note that $x_W = 1$, since it contains only water, $x_F = \frac{40}{100} = 0.4$, and $x_P = \frac{5}{100} = 0.05$). Hence:

$$800 \times 0.4 = (W \times 1) + P \times 0.05$$

$$320 = W + 0.05P \quad \text{...............Equation 4}$$

From equation 3:

$$W = 800 - P \quad \text{...............Equation 5}$$

Substitute 800 - P for W into equation 4. This gives:

$$320 = W + 0.05P \quad \text{...............Equation 4}$$

$$320 = (800 - P) + 0.05P$$

$$320 = 800 - 0.95P \quad \text{(Note that } P - 0.05P = 0.95P\text{)}$$

$$0.95P = 800 - 320$$

$$0.95P = 480$$

$$P = \frac{480}{0.95}$$

$$P = 505.3 \text{kg}$$

From equation 5:

$$W = 800 - P$$

$$= 800 - 505.3$$

$$W = 294.7 \text{kg}$$

Therefore the mass of water removed is 294.7kg.

This process can be summarized as follows:

In the feed: mass of water = 0.4 x 800 = 320kg

mass of flour = 800 = 320 = 480kg

In the product: mass of water in product = 0.05 x 505.3 = 25.3kg

mass of flour = 505.3 - 25.3

=480kg (This shows that the flour has not changed from feed)

Mass of water removed = 294.7kg

Note that x_F was converted to fraction as: $\frac{40}{100}$ = 0.4. Similarly, $x_P = \frac{5}{100}$ = 0.05. The percentage of water in water, x_W = 100%, which was converted to fraction as: $\frac{100}{100}$ = 1

2. A paper pulp containing 60% of moisture is passed through a dryer in order to remove 80% of the original moisture. Calculate:

(a). The mass of water removed per kg of the pulp

(b). The composition of the product

Solution

Let F = paper pulp (i.e. the feed)

W = moisture removed

P = product (i.e. dried pulp)

Also, let x_F = % moisture in feed = 60% = 0.6

x_W = % moisture in moisture = 100% (water only) = 1

x_P = % moisture in product

The overall component balance is given by:

F = W + P

1 = W + P ....Equation 1   (Note that F = 1kg since we are to calculate for per kg of pulp)

The moisture component balance is given by:

$$Fx_F = Wx_W + Px_P$$

$$1 \times 0.6 = (W \times 1) + Px_P$$

$$0.6 = W + Px_P \ldots\ldots\ldots\ldots\text{Equation 2}$$

From equation 1:

$$P = 1 - W \ldots\ldots\ldots\ldots\text{Equation 3}$$

Amount of moisture in feed = 60%

Therefore moisture in feed = 0.6 x 1 = 0.6kg

Since 80% of this moisture was removed, then:

Amount of moisture removed = 0.8 x 0.6 = 0.48kg

Therefore, W = 0.48kg

(b). From equation 3,

$$P = 1 - W$$

$$= 1 - 0.48$$

$$P = 0.52\text{kg}$$

Substituting the values of W and P into equation 2 gives:

$$0.6 = W + Px_P$$

$$0.6 = 0.48 + 0.52x_P$$

$$0.6 - 0.48 = 0.52x_P$$

$$0.52x_P = 0.12$$

$$x_P = \frac{0.12}{0.52}$$

$$x_P = 0.23$$

Recall that the amount of moisture in feed = 0.6kg.

Therefore, the amount of pulp in feed = 1 - 0.6 = 0.4kg

This amount of pulp in feed, also passes completely to the product. Therefore, amount of pulp in product is 0.4kg.

But, product, P = 0.52kg

Therefore amount of water in product, i.e. dried pulp = 0.52 - 0.4

= 0.12kg.

Therefore the composition of the product is:

% of pulp = $\frac{0.4}{0.52}$ x 100 = 76.9%

% of moisture = $\frac{0.12}{0.52}$ x 100 = 23.1%

3. 50m^3/h of a 5% sodium chloride solution of density 1200kg/m^3 is concentrated by evaporation to produce 40% brine. Calculate the rate of evaporation of water from the salt solution.

Solution

Let F = mass flow rate of the feed, in kg/h

W = water removed in kg/h

P = product i.e. brine in kg/h

Also, let, x_F = % salt in feed = 0.05   (Salt is the NaCl)

x_W = % salt in water = 0 (No salt in evaporated water)

x_P = % salt in product = 0.4

Let us convert the volumetric flow rate to mass flow rate

Recall that: Density = $\frac{mass}{volume}$

Therefore mass = density x volume

Similarly, mass flow rate = density x volumetric flow rate

= 1200 x 50

= 60,000kg/h

Therefore, F = 60,000kg/h

The overall component balance is given by:

F = W + P

60,000 = W + P ....Equation 1

The salt component balance is given by:

$Fx_F = Wx_W + Px_P$

60,000 x 0.05 = (W x 0) + (P x 0.4)

3000 = 0.4P

Therefore, $P = \dfrac{3000}{0.4}$

P = 7500kg/h

From equation 1:

60,000 = W + P

60,000 = W + 7500

W = 60,000 - 7500

W = 52,500kg/h

Therefore, the rate of evaporation of water is 52,500kg/h.

4. A fresh wood containing 45% of moisture is dried in an oven in order to remove 95% of its moisture content. Calculate:

(a). The mass of water remove per 100kg of the wood

(b). The mass of the dried wood

<u>Solution</u>

Let F = fresh wood (i.e. the feed)

W = moisture removed

P = product (i.e. dried pulp)

Also, let x_F = % moisture in feed = 45% = 0.45

x_W = % moisture in moisture = 100% (water only) = 1

x_P = % moisture in product

The overall component balance is given by:

F = W + P

100 = W + P ....Equation 1   (Note that F = 100kg)

The moisture component balance is given by:

$Fx_F = Wx_W + Px_P$

100 x 0.45 = (W x 1) + Px_P

45 = W + Px_P ....Equation 2

From equation 1:

P = 100 - W ....Equation 3

Amount of moisture in feed = 45%

Therefore moisture in feed = 0.45 x 100 = 45kg

Since 95% of this moisture was removed, then:

Amount of moisture removed = 0.95 x 45

= 42.75kg

Therefore, W = 42.75kg

(b). From equation 3,

P = 100 - W

= 100 - 42.75

P = 57.25kg

Therefore the mass of the dried wood is 57.25kg

5. 2000kg of a 12% sugar solution is concentrated by evaporation to produce a 70% sugar solution. Calculate the mass of water evaporated.

Solution

Let F = mass of the feed

   W = water removed

   P = product

Also, let, x_F = % sugar in feed = 0.12

   x_W = % sugar in water = 0 (No sugar in evaporated water)

   x_P = % sugar in product = 0.7

The overall component balance is given by:

   F = W + P

   2000 = W + P ....Equation 1

The sugar component balance is given by:

   $Fx_F = Wx_W + Px_P$

2000 x 0.12 = (W x 0) + (P x 0.7)

   240 = 0.7P

Therefore, $P = \dfrac{240}{0.7}$

   P = 342.9kg

From equation 1:

   2000 = W + P

   2000 = W + 342.9

   W = 2000 - 342.9

W = 1657.1kg

Therefore, the mass of water evaporated is 1657.1kg.

EXERCISE

1. 2500kg of wet flour is fed into a dryer. The wet flour contains 45% by mass of water, while the dry product is to contain 4% by mass of water. Calculate the mass of water removed by the dryer.

2. A fabric material containing 72% of moisture is passed through a dryer in order to remove 94% of the original moisture. Calculate:

(a). The mass of water removed per kg of the fabric

(b). The composition of the product

3. 200m^3/h of a 7% sodium chloride solution of density 1910kg/m^3 is concentrated by evaporation to produce 85% brine. Calculate the rate of evaporation of water from the salt solution.

4. A fresh wood containing 82% of moisture is dried in an oven in order to remove 88% of its moisture content. Calculate:

(a). The mass of water remove per 5kg of the wood

(b). The mass of the dried wood

5. 400kg of a 5% sugar solution is concentrated by evaporation to produce a 58% sugar solution. Calculate the mass of water evaporated.

CHAPTER 5
BALANCES INVOLVING MIXING OF SOLUTIONS

In mixing of solutions, solutions are usually fed into a process to produce one output, provided the solutions do not react together.

Examples

1. 14% by mass of tetraoxosulphate (VI) acid solution is added into a process tank. 210kg of 75% tetraoxosulphate (VI) acid solution is also added to the tank. If the final solution gives 30% of sulphuric acid, determine the mass of this final acid solution.

Solution

Let F_1 = 14% tetraoxosulphate (VI) acid, i.e. the dilute acid

F_2 = 75% tetraoxosulphate (VI) acid, i.e. the concentrated acid

P = Product (The final acid solution)

Also, let x_{F1} = % by mass of acid in F_1 = 0.14

x_{F2} = % by mass of acid in F_2 = 0.75

x_P = % by mass of acid in P = 0.3

Therefore, the overall component balance is given by:

$F_1 + F_2 = P$

$F_1 + 210 = P$ ....Equation 1

The tetraoxosulphate (VI) acid component balance is given by:

$F_1 x_{F1} + F_2 x_{F2} = P x_P$

$(F_1 \times 0.14) + (210 \times 0.75) = (P \times 0.3)$

$0.14F_1 + 157.5 = 0.3P$ ....Equation 2

From equation 1:

$F_1 = P - 210$

Substitute P - 210 for F_1 in equation 2. This gives:

$$0.14F_1 + 157.5 = 0.3P \quad \text{............Equation 2}$$

$$0.14(P - 210) + 157.5 = 0.3P$$

$$0.14P - 29.4 + 157.5 = 0.3P$$

$$157.5 - 29.4 = 0.3P - 0.14P$$

$$128.1 = 0.16P$$

$$P = \frac{128.1}{0.16}$$

$$P = 800.6 \text{kg}$$

Therefore the mass of this final acid solution is 800.6kg.

Note that the mass of tetraoxosulphate (VI) acid in this final solution is = Px_P = 800.6 x 0.3 = 240.2kg

2. A fuel oil seller mixes two types of oils each containing n-heptane and iso-octane. He decides to mix 800kg of 90% iso-octane which is the first oil, with 70% iso-octane which is the second oil. If this produces 85% iso-octane, calculate:

(a). The mass of the second oil added to the mixture

(b). The mass of iso-octane in the product

Solution

Let F_1 = mass of first oil

F_2 = mass of second oil

P = mass of product

Also let x_{F1} = % by mass of iso-octane in F_1 = 0.9

x_{F2} = % by mass of iso-octane in F_2 = 0.7

x_P = % by mass of iso-octane in P = 0.85

Therefore the overall component mass balance is given by:

$F_1 + F_2 = P$

$$800 + F_2 = P \quad \text{................Equation 1}$$

The iso-octane component balance is given by:

$$F_1 x_{F1} + F_2 x_{F2} = P x_P$$

$$(800 \times 0.9) + (F_2 \times 0.7) = P \times 0.85$$

$$720 + 0.7F_2 = 0.85P \quad \text{................Equation 2}$$

From equation 1:

$$P = 800 + F_2 \quad \text{................Equation 3}$$

Substituting $800 + F_2$ for P in equation 2 gives:

$$720 + 0.7F_2 = 0.85P \quad \text{................Equation 2}$$

$$720 + 0.7F_2 = 0.85(800 + F_2)$$

$$720 + 0.7F_2 = 680 + 0.85F_2$$

$$720 - 680 = 0.85F_2 - 0.7F_2$$

$$40 = 0.15F_2$$

$$F_2 = \frac{40}{0.15}$$

$$F_2 = 266.7$$

The mass of the second oil in the mixture 266.7kg

(b). From equation 3 above:

$$P = 800 + F_2 \quad \text{................Equation 3}$$

$$= 800 + 266.7$$

$$P = 1066.7 \text{kg}$$

But mass of iso-octane in the product is given by:

$$P x_P$$

$$= 1066.7 \times 0.85$$

$$= 906.7 \text{kg}$$

Therefore mass of iso-octane in the product is 906.7kg.

3. A fruit juice is made from a mixture of juice and water. A fruit juice containing 20% of juice and 80% of water sells for $90/kg. It is mixed with a fruit juice containing 40% of juice which sells for $50/kg. If the fruit juice produced contains 34% of juice, how much should it be sold to make a profit of 20%

Solution

The prices of the fruit juice are the cost of 1kg of the fruit juice.

So, let us solve this problem by taking a basis of 1kg of product produced.

Let F_1 = mass of 20% juice

F_2 = mass of 40% juice

P = mass of product = 1kg

Also, let x_1 = % by mass of juice in F_1 = 0.20

x_2 = % by mass of juice in F_2 = 0.4

x_P = % by mass of juice in P = 0.34

The overall component balance is given by:

$F_1 + F_2 = P$

$F_1 + F_2 = 1$ (Since P = 1kg)

Or, $F_1 = 1 - F_2$ ....Equation 1

The juice component balance is given by:

$F_1x_1 + F_2x_2 = Px_P$

$0.20F_1 + 0.4F_2 = 1 \times 0.34$

$0.2F_1 + 0.4F_2 = 0.34$ ....Equation 2

Substitute $1 - F_2$ for F_1 from equation 1 into equation 2. This gives:

$0.2F_1 + 0.4F_2 = 0.34$ ....Equation 2

$0.2(1 - F_2) + 0.4F_2 = 0.34$

$0.2 - 0.2F_2 + 0.4F_2 = 0.34$

$0.4F_2 - 0.2F_2 = 0.34 - 0.2$

$0.2F_2 = 0.14$

$F_2 = \dfrac{0.14}{0.2}$

$F_2 = 0.7$ kg

From equation 1:

$F_1 = 1 - F_2$

$= 1 - 0.7$

$F_1 = 0.3$ kg

From the question, 1 kg of F_1 cost $90 while 1kg of F_2 cost $50. Therefore, 0.3kg of F_1 and 0.7kg of F_2 will cost:

$(0.3 \times 90) + (0.7 \times 50)$

$= 27 + 35$

$= \$62$

In order to make 20% profit on this cost, the juice must be sold for:

120% of cost price   [since 20% profit means 120% (i.e. 100 + 20 = 120) of cost price]

$= \dfrac{120}{100} \times 62$

$= 74.4$

The fruit juice should be sold for $74.40

4. 400 litres of 80% by mass of HCl is to be obtained by mixing two solutions of HCl. If this is obtained by mixing 90% of HCl and 50% of HCl solutions, calculate the volume (in litres) of each of these two solutions that are mixed together.

(Density of 80% HCl = 1.34kg/litre, Density of 90% HCl = 1.42kg/litre, Density of 50% HCl = 1.25kg/litre)

Solution

Since mass fractions are given, the volume of 400 litres must be converted to mass.

Recall that, density = $\dfrac{mass}{volume}$

Therefore, mass = density x volume

Hence, mass of 400 litres, 80% HCl = 1.34 x 400   (Its density is 1.34)

$\qquad\qquad\qquad\qquad$ = 536kg

Hence, the product, P = 536kg

Let: F_1 = 90% HCl

$\qquad F_2$ = 50% HCl

$\qquad x_1$ = 0.9

$\qquad x_2$ = 0.5

$\qquad x_P$ = 0.8   (Note that the 400 litres 80% HCl is the product)

The overall component mass balance is given by:

$\qquad F_1 + F_2 = P$

$\qquad F_1 + F_2 = 536$ ....Equation 1

The HCl component balance is given by:

$\qquad F_1x_1 + F_2x_2 = Px_P$

$(F_1$ x 0.9) + $(F_2$ x 0.5) = 536 x 0.8

$\qquad\qquad 0.9F_1 + 0.5F_2 = 428.8$ ....Equation 2

From equation 1:

$\qquad F_1 = 536 - F_2$ ....Equation 3

Substitute 536 - F_2 for F_1 in equation 2

$$0.9F_1 + 0.5F_2 = 428.8 \dots\dots\dots\dots\dots\text{Equation 2}$$

$$0.9(536 - F_2) + 0.5F_2 = 428.8$$

$$482.4 - 0.9F_2 + 0.5F_2 = 428.8$$

$$482.4 - 428.8 = 0.9F_2 - 0.5F_2$$

$$53.6 = 0.4F_2$$

$$F_2 = \frac{53.6}{0.4}$$

$$F_2 = 134\text{kg}$$

From equation 3:

$$F_1 = 536 - F_2 \dots\dots\dots\dots\text{Equation 3}$$

$$= 536 - 134$$

$$F_1 = 402\text{kg}$$

In order to obtain F_1 and F_2 in litres (volume), their densities have to be applied.

Recall that, density $= \dfrac{\text{mass}}{\text{volume}}$

Therefore, volume $= \dfrac{\text{mass}}{\text{density}}$

Hence, volume of F_1 in litres $= \dfrac{402}{1.42}$ (F_1 is 90% HCl and has density of 1.42)

$$= 283.8 \text{ litres}$$

Volume of F_2 in litres $= \dfrac{134}{1.25}$ (F_2 = 50% HCl and has a density of 1.25)

$$= 107.2 \text{ litres}$$

5. A paint seller mixes two types of paints. A paint containing 30% of pigment sells for $20/kg. It is mixed with a paint containing 65% of pigment which sells for $52/kg. If the paint produced contains 46% of pigment, how much should it be sold?

Solution

The prices of the paint are the cost of 1kg of the paint.

So, let us solve this problem by taking 1kg of product formed.

Let F_1 = mass of 30% pigment

F_2 = mass of 65% pigment

P = mass of product = 1kg

Also, let x_1 = % by mass of pigment in F_1 = 0.30

x_2 = % by mass of pigment in F_2 = 0.65

x_P = % by mass of pigment in P = 0.46

The overall component balance is given by:

$F_1 + F_2 = P$

$F_1 + F_2 = 1$ (Since P = 1kg)

Or, $F_1 = 1 - F_2$ ....Equation 1

The pigment component balance is given by:

$F_1x_1 + F_2x_2 = Px_P$

$0.30F_1 + 0.65F_2 = 1 \times 0.46$

$0.3F_1 + 0.65F_2 = 0.46$ ....Equation 2

Substitute $1 - F_2$ for F_1 from equation 1 into equation 2. This gives:

$0.3F_1 + 0.65F_2 = 0.46$ ....Equation 2

$0.3(1 - F_2) + 0.65F_2 = 0.46$

$0.3 - 0.3F_2 + 0.65F_2 = 0.46$

$0.65F_2 - 0.3F_2 = 0.46 - 0.3$

$0.35F_2 = 0.16$

$$F_2 = \frac{0.16}{0.35}$$

$F_2 = 0.457$kg

From equation 1:

$$F_1 = 1 - F_2$$
$$= 1 - 0.457$$
$$F_1 = 0.543 \text{ kg}$$

From the question, 1 kg of F_1 cost $20 while 1kg of F_2 cost $52. Therefore, 0.543kg of F_1 and 0.457kg of F_2 will cost:

$$(0.543 \times 20) + (0.457 \times 52)$$
$$= 10.86 + 23.76$$
$$= \$34.62$$

Hence the paint should be sold for $34.62

EXERCISE

1. 10% by mass of tetraoxosulphate (VI) acid solution is added into a process tank. 500kg of 50% tetraoxosulphate (VI) acid solution is also added to the tank. If the final solution gives 24% of sulphuric acid, determine the mass of this final acid solution.

2. A fuel oil seller mixes two types of oils each containing n-heptane and iso-octane. He decides to mix 1500kg of 80% iso-octane which is the first oil, with 35% iso-octane which is the second oil. If this produces 66% iso-octane, calculate:

(a). The mass of the second oil added to the mixture

(b). The mass of iso-octane in the product

3. A fruit juice is made from a mixture of juice and water. A fruit juice containing 18% of juice and 82% of water sells for $60/kg. It is mixed with a fruit juice containing 45% of juice which sells for $34/kg. If the fruit juice produced contains 40% of juice, how much should it be sold to make a profit of 45%

4. 20 litres of 90% by mass of HCl is to be obtained by mixing two solutions of HCl. If this is obtained by mixing 98% of HCl and 64% of HCl solutions, calculate the volume (in litres) of each of these two solutions that are mixed together.

(Density of 90% HCl = 1.15kg/litre, Density of 98% HCl = 1.28kg/litre, Density of 64% HCl = 1.10kg/litre)

5. A paint seller mixes two types of paints. A paint containing 50% of pigment sells for $12/kg. It is mixed with a paint containing 22% of pigment which sells for $5/kg. If the paint produced contains 36% of pigment, how much should it be sold?

CHAPTER 6
BALANCES INVOLVING COMBUSTION

Some of the terms associated with combustion are:

(I). Stack or flue gas: This involves all the gases that result from a combustion process. Flue or stack gas contains water vapour, hence it is a wet gas.

(II). Orsat analysis: This is a dry gas analysis because it contains all the gases excluding water vapour from a combustion process.

In combustion process calculations, excess air or oxygen and theoretical air or oxygen are usually calculated.

Examples

1. 20kg of methane is combusted with 380kg of air to give 36kg of carbon (IV) oxide and 14kg of carbon (II) oxide. Determine the percent excess air used.

(Air contains 23.2% by mass of oxygen, and the molecular mass of air is 29kg/kmol)

Solution

Since excess air was supplied, it means that the reaction is a complete combustion. The first step is to calculate the amount of oxygen needed for the combustion. This is obtained from the complete combustion reaction.

The balanced equation for the complete combustion reaction is:

$$CH_4 + 2O_2 \longrightarrow CO_2 + 2H_2O$$
$$16kg + 64kg \longrightarrow 44kg + 36kg$$

Note that the molecular mass of CH_4 = 16, the molecular mass of 2kmoles of O_2 = 2 x 32 = 64, the molecular mass of CO_2 = 44, while the molecular mass of 2kmoles of H_2O = 2 x 18 = 36. These are the masses shown above. The above equation shows that:

16kg of CH_4 requires 64kg of O_2 for complete combustion. Therefore, by simple proportion 20kg (from the question) of CH_4 will require:

$$\frac{20}{16} \times 64 \quad \text{of } O_2$$

= 80kg of O_2

From the question, air supplied = 380kg. But air contains 23.2% by mass of O_2

Therefore by simple proportion, mass of O_2 in air supplied = $\frac{23.2}{100}$ x 380

= 0.232 x 380

= 88.16kg

Hence, % excess air used = % excess O_2 used

% excess air used = $\frac{O_2 \text{ supplied } - O_2 \text{ required}}{O_2 \text{ required}}$ x 100

= $\frac{88.16 - 80}{80}$ x 100

= 10.2%

Therefore, percent excess air supplied is 10.2%

2. During a combustion process 32 litres of ethene is burnt with 540 litres of air to give 50 litres of carbon (IV) oxide and 18 litres of carbon (II) oxide. Calculate the percent excess air supplied.

(Air contains 21% by volume of oxygen)

Solution

The reaction is a complete combustion reaction since excess air was supplied. The volumes given in the question can be used as moles or kmoles.

The balanced equation for the reaction is:

C_2H_4 + $3O_2$ ----------> $2CO_2$ + $2H_2O$
1kmol + 3kmol -------> 2kmol + 2kmol

This balanced equation shows that 1kmol of C_2H_4 requires 3kmol of O_2 for complete combustion. Therefore, by simple proportion, 32 litres of C_2H_4 will require:

$\frac{32}{1}$ x 3 of O_2 (Note that moles are used as volumes in reactions)

= 96 litres of O_2

However, air supplied = 540 litres. But air contains 21% by volume of O_2

Therefore, by simple proportion, volume of O_2 supplied = $\frac{21}{100}$ x 540

= 0.21 x 540

= 113.4 litres

Therefore, % excess air = $\frac{O_2 \text{ supplied } - O_2 \text{ required}}{O_2 \text{ required}}$ x 100

= $\frac{113.4 - 96}{96}$ x 100

= 18.1%

3. A gas has the following composition by volume:
5% of CO_2, 12% of CO, 3.8% of C_2H_2, 18% of C_3H_8, 55% of H_2, 0.2% of O_2, and 6% of N_2. The gas is burnt to produce a stack gas containing the following composition by volume on a dry basis: 7.0% of CO_2, 7.6% of O_2, and 85.4% of N_2. Calculate the required and actual air supplied.

Solution

The components of the gas that will undergo combustion are CO, C_2H_4, C_3H_8 and H_2. CO_2, O_2 and N_2 will pass directly to the product stream.

Let us use a basis of 100kmoles of feed.

The combustion reactions for each of these four components that will undergo combustion are as given below.

$$CO + ½O_2 \longrightarrow CO_2$$

$$12CO + (12 \times ½)O_2 \longrightarrow 12CO_2 \quad (12\% \text{ of CO represent 12kmol})$$

$$12CO + 6O_2 \longrightarrow 12CO_2 \quad \text{..................Equation 1}$$

$$C_2H_2 + 3O_2 \longrightarrow 2CO_2 + 2H_2O$$

$3.8C_2H_2 + (3.8 \times 3)O_2 \longrightarrow (3.8 \times 2)CO_2 + (3.8 \times 2)H_2O$ (from the 3.8% of C_2H_4 in feed)

$3.8C_2H_2 + 11.4O_2 \longrightarrow 7.6CO_2 + 7.6H_2O$ ....Equation 2

$C_3H_8 + 5O_2 \longrightarrow 3CO_2 + 4H_2O$

$18C_3H_8 + (18 \times 5)O_2 \longrightarrow (18 \times 3)CO_2 + (18 \times 4)H_2O$

$18C_3H_8 + 90O_2 \longrightarrow 54CO_2 + 72H_2O$ ....Equation 3

$55H_2 + (55 \times ½)O_2 \longrightarrow 55H_2O$

$55H_2 + 27.5O_2 \longrightarrow 55H_2O$ ....Equation 4

Bringing equations 1 to 4 together gives:

$12CO + 6O_2 \longrightarrow 12CO_2$ ....Equation 1

$3.8C_2H_2 + 11.4O_2 \longrightarrow 7.6CO_2 + 7.6H_2O$ ....Equation 2

$18C_3H_8 + 90O_2 \longrightarrow 54CO_2 + 72H_2O$ ....Equation 3

$55H_2 + 27.5O_2 \longrightarrow 55H_2O$ ....Equation 4

From these four equations the total kmol of O_2 required = 6 + 11.4 + 90 + 27.5

    = 134.9 kmols

Total kmol of CO_2 = 12 + 7.6 + 54

    = 73.6

But the 5% of CO_2 in the reactant (from the question) did not react but passed directly to the product. So this is added to the value above as follows:

Total kmol of CO_2 in product = 73.6 + 5

    = 78.6

Total kmol of H_2O in product = 7.6 + 72 + 55

    = 134.6

Recall that air contains 21% by volume of O_2

Therefore the required O_2 of 134.9kmol represents 21%

Hence, 100% air will give: $\dfrac{100}{21}$ x 134.9

= 642.8kmol of air

Therefore the theoretical air required is 642.8kmol.

The N_2 in air from the feed will pass directly to the product. Hence, let us calculate the actual air supplied by using the kmol of N_2 in the product (this came from the feed of air).

Recall that the kmol of CO_2 in the product = 78.6. This 78.6kmol represent 7% (i.e. % of CO_2 in the product).

Therefore, by simple proportion, 100% product will be = $\dfrac{100}{7}$ x 78.6

= 1123kmol

Hence total kmol of product is 1123kmol

Since 85.4% of N_2 is present in the product, then kmol of N_2 in product = $\dfrac{85.4}{100}$ x 1123

= 959kmol

This 959kmol of N_2 contains the 6% of N_2 from the feed that passed directly to the product stream.

Therefore kmol of N_2 from air = 959 - 6

= 953kmol

Recall that air contains 79% by volume of N_2.

Therefore, 79% N_2 represent 953kmol

Hence, 100% air will be = $\dfrac{100}{79}$ x 953

= 1206kmol

Therefore actual air supplied is 1206kmols of air.

Another method of solving this problem is to use the individual element, C and H_2 in the hydrocarbons directly.

Therefore the C_2H_4 can be broken into its elements as follows.

$$C_2H_4 \longrightarrow 2C + 2H_2$$

Thus, $3.8C_2H_4 \longrightarrow (3.8 \times 2)C + (3.8 \times 2)H_2$

$$3.8C_2H_4 \longrightarrow 7.6C + 7.6H_2 \quad \text{..............Equation 5}$$

Similarly, C_3H_8 can be broken into its component elements as follows.

$$C_3H_8 \longrightarrow 3C + 4H_2$$

Thus, $18C_3H_8 \longrightarrow (18 \times 3)C + (18 \times 4)H_2$

Or, $18C_3H_8 \longrightarrow 54C + 72H_2 \quad \text{..............Equation 6}$

Hence the total kmol of C in the hydrocarbons from equations 5 and 6 is:

$7.6 + 54 = 61.6$ kmol

The total kmol of H_2 from equations 5 and 6 is:

$7.6 + 72 = 79.6$ kmols

These total kmol of C and H_2 can now be used to write simple combustion equations as follows.

$$C + O_2 \longrightarrow CO_2$$

$$61.6C + 61.6O_2 \longrightarrow 61.6CO_2 \quad \text{..............Equation 7}$$

For the H_2 we have:

$$H_2 + \tfrac{1}{2}O_2 \longrightarrow H_2O$$
$$79.6H_2 + (79.6 \times \tfrac{1}{2})O_2 \longrightarrow 79.6H_2O$$

$$79.6H_2 + 39.8O_2 \longrightarrow 79.6H_2O \quad \text{..............Equation 8}$$

Recall that equations 1 and 4 are:

$$12CO + 6O_2 \longrightarrow 12CO_2 \quad \text{..............Equation 1}$$

$$55H_2 + 27.5O_2 \longrightarrow 55H_2O \quad \text{..............Equation 4}$$

Hence, by using equations 7, 8, 1 and 4, we have:

Total kmol of O_2 required = 61.6 + 39.8 + 6 + 27.5

$$= 134.9 \text{kmols}$$

Total kmol of CO_2 in product = 61.6 + 12 + 5 (i.e. CO_2 from feed)

$$= 78.6 \text{kmol of } CO_2$$

Total kmol of H_2O in product = 79.6 + 55

$$= 134.6 \text{ kmol of } H_2O$$

Therefore, these three values of 134.9kmol of O_2, 78.6 kmol of CO_2 and 134.6kmol of H_2O which are the same as obtained in the first method, can now be used to calculate the required and actual air supplied as carried out in the first method.

4. A fuel which contains carbon and hydrogen only, was burnt to produce a dry flue gas with the following composition by volume: 9.6% of CO_2, 8.4% of O_2 and 82.0% of N_2. Calculate:

(a). The composition of the fuel

(b). The percent excess air supplied

Solution

(a). A basis of 100moles of flue gas (product) will be used. This means that the volume of CO_2 produced is 9.6moles, the volume of O_2 produced is 8.4moles, while the volume of N_2 in the product is 82moles. So, we will work from the product back to the reactant.

The combustion reaction for carbon is:

$$C + O_2 \longrightarrow CO_2$$

Since 9.6moles of CO_2 was produced, by working backward, the moles of C and O_2 are obtained as follows:

$$9.6C + 9.6O_2 \longrightarrow 9.6CO_2 \quad \text{...................Equation 1}$$

The N_2 did not undergo any reaction. So, all the 82moles of N_2 came from the air. Recall that air contains 79% by volume of N_2. So by simple proportion:

79% N_2 gives 82moles

Therefore, 100% air will give: $\frac{100}{79} \times 82$

$= 103.8 moles$

Therefore, the moles of air supplied is 103.8moles

But air contains 21% by volume of O_2. Therefore, total O_2 supplied from the air is:

$\frac{21}{100} \times 103.8$

$= 0.21 \times 103.8$

$= 21.8 moles$

From the question 8.4moles of O_2 was present in the product stream. This is the amount of excess O_2 that did not take part in the reaction.

Therefore moles of O_2 that reacted = 21.8 - 8.4

$= 13.4 moles$

Out of this 13.4moles, 9.6moles from equation 1 has reacted with C. Therefore, the remaining O_2 will react with H_2.

Therefore O_2 that react with H_2 = 13.4 - 9.6

$= 3.8 moles$

Hence the H_2 combustion reaction is represented as:

$H_2 + 3.8O_2 \longrightarrow H_2O$

In order to balance this reaction, we have to make the oxygen atoms on both sides of the reaction to be 2 x 3.8 (since this is total O_2 atoms on the left). This gives:

$(2 \times 3.8)H_2 + 3.8O_2 \longrightarrow (2 \times 3.8)H_2O$

Or, $7.6H_2 + 3.8O_2 \longrightarrow 7.6H_2O$ ....Equation 2

Therefore, from equation 1, the moles of C = 9.6moles, and from equation 2 the moles of H_2 = 7.6moles.

Total moles of C and H_2 = 9.6 + 7.6

$= 17.2 moles$

Hence, the composition of C and H_2 in the fuel, are:

$$\% \text{ composition of C} = \frac{9.6}{17.2} \times 100$$

$$= 55.8\%$$

$$\% \text{ composition of } H_2 = \frac{7.6}{17.2} \times 100$$

$$= 44.2.1\%$$

(b). Recall that the total O_2 supplied = 21.8moles, and the unreacted (excess) O_2 = 8.4moles.

Therefore, the reacted (used) O_2 = 21.8 - 8.4

$$= 13.4 \text{moles}$$

Therefore, % excess air supplied $= \dfrac{\text{supplied } O_2 - \text{required } O_2}{\text{required } O_2} \times 100$

$$= \frac{\text{excess } O_2}{\text{required } O_2} \times 100$$

$$= \frac{8.4}{13.4} \times 100$$

$$= 62.7\%$$

5. A petroleum gas was analyzed to contain the following composition by volume: 74% of C_3H_8, 22.6% of C_4H_{10}, and 3.4% of CO_2. The gas was burnt with 28% excess air. 92% of the hydrocarbons is converted to CO_2 while 8% is converted to CO. Calculate the composition of the flue gas.

Solution

Let us work on a basis of 100 moles of the petroleum gas. Note that when hydrocarbons burn, the products formed from complete combustion are CO_2 and H_2O. So, the product stream for this equation will contain CO_2, H_2O, CO (from incomplete combustion), O_2 (from excess air) and N_2 (from air).

For complete combustion of each of the hydrocarbons we have:

$$C_3H_8 + 5O_2 \longrightarrow 3CO_2 + 4H_2O$$

Using the 74moles (i.e. 74% of C_3H_8), gives:

$$74C_3H_8 + (74 \times 5)O_2 \longrightarrow (74 \times 3)CO_2 + (74 \times 4)H_2O$$

Or, $\quad 74C_3H_8 + 370O_2 \longrightarrow 222CO_2 + 296H_2O$ ....Equation 1

And, $\quad C_4H_{10} + \dfrac{13}{2}O_2 \longrightarrow 4CO_2 + 5H_2O$

Using the 22.6moles of C_4H_{10} gives:

$$22.6C_4H_{10} + (22.6 \times \dfrac{13}{2})O_2 \longrightarrow (22.6 \times 4)CO_2 + (22.6 \times 5)H_2O$$

Or, $22.6C_4H_{10} + 146.9O_2 \longrightarrow 90.4CO_2 + 113H_2O$ ....Equation 2

Therefore the total O_2 for complete combustion from equations 1 and 2 is:

$\quad 370 + 146.9 = 516.9$ moles

Since 28% excess air was supplied, then the total O_2 supplied is given by:

$\quad \dfrac{128}{100} \times 516.9 \quad$ (28% of excess air means 128% (i.e. 100 + 28) of required air)

$\quad = 1.28 \times 516.9$

$\quad = 661.63$ moles

This amount of O_2 corresponds to 21% by volume of O_2 in air. Therefore, the 79% by volume of N_2 associated with this O_2 will be obtained as follows:

$\quad$ 21% of O_2 gives 661.63moles

Therefore, 79% of N_2 will give: $\dfrac{79}{21} \times 661.63$

$\quad\quad\quad\quad\quad\quad\quad\quad\quad = 2489.0$ moles of N_2

From the question, the hydrocarbons are also converted to CO as follows:

$$C_3H_8 + \dfrac{7}{2}O_2 \longrightarrow 3CO + 4H_2O$$

Using the 74moles of C_3H_8 gives:

$$74C_3H_8 + (74 \times \dfrac{7}{2})O_2 \longrightarrow (74 \times 3)CO + (74 \times 4)H_2O$$

Or, $\quad 74C_3H_8 + 259O_2 \longrightarrow 222CO + 296H_2O$ ....Equation 3

And, $\quad C_4H_{10} + \dfrac{9}{2}O_2 \text{----------} > 4CO + 5H_2O$

Using the 22.6 moles of C_4H_{10} gives:

$\quad 22.6C_4H_{10} + (22.6 \times \dfrac{9}{2})O_2 \text{----------} > (22.6 \times 4)CO + (22.6 \times 5)H_2O$

Or, $\quad 22.6C_4H_{10} + 101.7O_2 \text{----------} > 90.4CO + 113H_2O$ ....Equation 4

However, using equations 1 and 2, 92% conversion of the hydrocarbons to CO_2 gives:

$\quad 0.92(74C_3H_8 + 370O_2 \text{----------} > 222CO_2 + 296H_2O)$

Or, $\quad 68.08C_3H_8 + 340.4O_2 \text{----------} > 204.24CO_2 + 272.32H_2O$ ....Equation 5

And, $0.92(22.6C_4H_{10} + 146.9O_2 \text{----------} > 90.4CO_2 + 113H_2O)$

Or, $\quad 20.79C_4H_{10} + 135.15O_2 \text{----------} > 83.17CO_2 + 103.96H_2O$ ....Equation 6

Similarly, from equations 3 and 4, 8% conversion of hydrocarbon to CO gives:

$\quad 0.08(74C_3H_8 + 259O_2 \text{----------} > 222CO + 296H_2O)$

Or, $\quad 5.92C_3H_8 + 20.72O_2 \text{----------} > 17.76CO + 23.68H_2O$ ....Equation 7

And, $0.08(22.6C_4H_{10} + 101.7O_2 \text{----------} > 90.4CO + 113H_2O)$

Or, $\quad 1.81C_4H_{10} + 8.14O_2 \text{----------} > 7.26CO + 9.04H_2O$ ....Equation 8

From equations 5 to 8, the total O_2 used is:

$\quad 340.4 + 135.15 + 20.72 + 8.14 = 504.41 \text{moles}$

Total CO_2 produced from equations 5 and 6 is:

$\quad 204.24 + 83.18 = 287.41 \text{moles}$

But total CO_2 in product = 287.41 + 3.4 (i.e. CO_2 from petroleum gas is 3.4)

$\quad\quad\quad = 290.81 \text{moles}$

Total H_2O from equations 1 and 2 (i.e. complete combustion) is:

$\quad 296 + 113 = 409 \text{moles}$

This can also be obtained from equations 5 to 8.

Total CO from equations 7 and 8 is:

$$17.76 + 7.26 = 25.02 \text{ moles}$$

Total N_2 in product remains 2489 moles

Total O_2 in product (i.e. unreacted or excess O_2) is:

O_2 from excess air supplied - O_2 that reacted

$$= 661.63 - 504.41$$

$$= 157.22 \text{ moles}$$

Therefore, total moles of all gases in the product, is:

157.22 moles of O_2 + 2489 moles of N_2 + 25.02 moles of CO + 409 moles of H_2O + 290.81 moles of CO_2 = 3371.05 moles

Therefore the composition of the flue gas is:

$$\% \ O_2 = \frac{157.22}{3371.05} \times 100$$

$$= 4.66\%$$

$$\% \ N_2 = \frac{2489}{3371.05} \times 100$$

$$= 73.83\%$$

$$\% \ CO = \frac{25.02}{3371.05} \times 100$$

$$= 0.74\%$$

$$\% \ H_2O = \frac{409}{3371.05} \times 100$$

$$= 12.13\%$$

$$\% \ CO_2 = \frac{290.81}{3371.05} \times 100$$

$$= 8.63\%$$

6. A crude oil was analyzed to contain the following composition by mass: 75% of C, 19.4% of H_2 and 5.6% of S. If the stack gas is produced at 250°C and a pressure of $1.04 \times 10^5 \text{ N/m}^2$, calculate:

(a). The composition of the stack gas from combustion of 100kg of the crude oil

(b). The volume of the stack gas.

Solution

(a). Note that the stack gas will contain CO_2 from C, H_2O from H_2, SO_2 from S and N_2 from air used for the combustion.

The complete combustion reaction for C is:

$$C + O_2 \longrightarrow CO_2$$

Using the molecular mass of each substance shows that:

$$12\text{kgC} + 32\text{kgO}_2 \longrightarrow 44\text{kgCO}_2$$

By dividing each substance by 12kg, then 1kg of C will react as follows:

$$1\text{kgC} + \frac{32}{12}\text{kgO}_2 \longrightarrow \frac{44}{12}\text{kgCO}_2$$

Therefore 75kg of C (from question) will react as follows:

$$(75 \times 1)C + (75 \times \frac{32}{12})O_2 \longrightarrow (75 \times \frac{44}{12})CO_2$$

Or, $75C + 200O_2 \longrightarrow 275CO_2$ ....Equation 1

The reaction of H_2 is given by:

$$H_2 + \tfrac{1}{2}O_2 \longrightarrow H_2O$$

Hence, $2\text{kgH}_2 + (32 \times \tfrac{1}{2})\text{kgO}_2 \longrightarrow 18\text{kgH}_2O$

Or, $2\text{kgH}_2 + 16\text{kgO}_2 \longrightarrow 18\text{kgH}_2O$

Hence, by dividing each substance by 2, it means that 1kg of H_2 will react as follows:

$$1\text{kgH}_2 + \frac{16}{2}\text{kgO}_2 \longrightarrow \frac{18}{2}\text{kgH}_2O$$

Or, $1\text{kgH}_2 + 8\text{kgO}_2 \longrightarrow 9\text{kgH}_2O$

Therefore the 19.4kg of H_2 in the crude oil will react as follows:

$$(19.4 \times 1)H_2 + (19.4 \times 8)O_2 \longrightarrow (19.4 \times 9)H_2O$$

Or, $19.4H_2 + 155.2O_2 \longrightarrow 174.6H_2O$ ....Equation 2

The reaction for S is given by:

$$S + O_2 \longrightarrow SO_2$$

$32kgS + 32kgO_2 \longrightarrow 64kgSO_2$ (Note that the molecular mass of S is 32)

Hence by dividing each substance by 32, it means that 1kg of S will react as follows:

$$1kgS + \frac{32}{32}kgO_2 \longrightarrow \frac{64}{32}kgSO_2$$

Or, $1kgS + 1kgO_2 \longrightarrow 2kgSO_2$

Therefore the 5.6kg of S in the crude oil will react as follows:

$$(5.6 \times 1)S + (5.6 \times 1)O_2 \longrightarrow (5.6 \times 2)SO_2$$

Or, $5.6S + 5.6O_2 \longrightarrow 11.2SO_2$ ....Equation 3

Therefore from equation 1 to 3, the mass of O_2 required for complete combustion is:

$200 + 155.2 + 5.6 = 360.8kg$

Recall that air contains 23.2% by mass of O_2. So, by simple proportion, the mass of N_2 in the air can be obtained as follows, and noting that air contains 76.8% by mass of N_2.

23.2% gives 360.8kg of O_2

Therefore, 76.8% will give: $\frac{76.8}{23.2} \times 360.8$ of N_2

$= 1194.4kg$ of N_2

Therefore, mass of N_2 in product is 1194.4kg

Mass of CO_2 in product (from equation 1) is 275kg

Mass of H_2O in product from (from equation 2) is 174.6kg

Mass of SO_2 in product (from equation 3) is 11.2kg

Hence the total mass of the components in the product is:

1194.4 + 275 + 174.6 + 11.2 = 1655.2kg

Therefore, the composition of the stack gas is:

% of N_2 = $\dfrac{1194.4}{1655.2}$ x 100

= 72.2%

% of CO_2 = $\dfrac{275}{1655.2}$ x 100

= 16.6%

% of H_2O = $\dfrac{174.6}{1655.2}$ x 100

= 10.5%

% of SO_2 = $\dfrac{11.2}{1655.2}$ x 100

= 0.7%

(b). Converting each of the components of the product to kmols gives the following, and noting that: Number of moles = $\dfrac{mass}{molecular\ mass}$

Therefore, Kmols of N_2 = $\dfrac{1194.4}{28}$

= 42.68kmols

Kmols of CO_2 = $\dfrac{275}{44}$

= 6.25kmols

kmols of H_2O = $\dfrac{174.6}{18}$

= 9.7kmols

Kmols of SO_2 = $\dfrac{11.2}{64}$

= 0.18

Hence, total kmols of stack gas = 42.68 + 6.25 + 9.7 + 0.18

= 58.81 kmols

Recall that 1 kmol of any gas occupies a volume of $22.4m^3$. Therefore the volume occupied by 58.81 kmols of stack gas is:

$$58.81 \times 22.4 = 1317.3 m^3$$

Therefore, by using the general gas law, the initial conditions of the stack gas at s.t.p are:

$P_1 = 1.01 \times 10^5 N/m^2$, $T_1 = 273k$, $V_1 = 1317.3 m^3$. The final conditions of the stack gas are:

$P_2 = 1.04 \times 10^5 N/m^2$, $T_2 = 250 + 273 = 523k$, $V_2 = ?$ (Note that 250°C is given in the question)

Hence, from the general gas law:

$$\frac{P_1 V_1}{T_1} = \frac{P_2 V_2}{T_2}$$

Therefore, $V_2 = \dfrac{P_1 V_1 T_2}{P_2 T_1}$

$$= \frac{1.01 \times 1317.3 \times 523}{1.04 \times 273}$$ (Note that 10^5 from pressure has canceled out)

$V_2 = 2450.8 m^3$

7. Ethane is burnt with 20% excess air. Fractional conversion is 84%. Calculate the percentage composition of each of the components of the flue gas.

Solution

For complete combustion the equation below follows:

$$C_2H_6 + \frac{7}{2} O_2 \longrightarrow 2CO_2 + 3H_2O$$

Working on a basis of 1 mole of ethane as the feed, the equation above shows that:

$\frac{7}{2}$ or 3.5 moles of O_2 is required

But, 20% excess air was supplied.

Therefore, moles of O_2 supplied $= \dfrac{120}{100} \times 3.5$

$$= 1.2 \times 3.5$$

$$= 4.2 \text{ moles}$$

Air contains 21% O_2 and 79% N_2 by volume. So, the moles of N_2 in the air supplied, is obtained as follows:

21% O_2 gives 4.2 moles

Therefore, 79% will give: $\frac{79}{21} \times 4.2$

$$= 15.8 \text{ moles of } N_2$$

This shows that the moles of N_2 in the product is 15.8 moles

Now, for the 84% fractional conversion, the above reaction equation is given as:

$$0.84(C_2H_6 + 3.5O_2 \text{----------> } 2CO_2 + 3H_2O)$$

Or, $\quad 0.84C_2H_6 + 2.94O_2 \text{----------> } 1.68CO_2 + 2.52H_2O$ ....Equation 1

It follows from this reaction that 0.84 moles of C_2H_6 reacted. So, moles of C_2H_6 that is in product (i.e. unreacted C_2H_6) is given by:

$$= 1 - 0.84 \quad \text{(Note that feed is 1 mole of } C_2H_6\text{)}$$

$$= 0.16 \text{ moles}$$

From equation 1 above, 2.94 moles of O_2 reacted. So moles of O_2 that is in the product (i.e. unreacted O_2) = 4.2 - 2.94

$$= 1.26 \text{ moles}$$

Moles of CO_2 produced, i.e. CO_2 in product = 1.68 moles

Moles of H_2O in product = 2.52 moles   (All from equation 1)

Thus the flue gas contains:

15.8 moles of N_2

0.16 moles of C_2H_6

1.26 moles of O_2

1.68 moles of CO_2

2.52 moles of H$_2$O

Total moles = 21.42   (When all the values above are added)

Therefore the flue gas contains the following percentage composition:

$$\% \text{ N}_2 = \frac{15.8}{21.42} \times 100$$

$$= 73.8\%$$

$$\% \text{ C}_2\text{H}_6 = \frac{0.16}{21.42} \times 100$$

$$= 0.7\%$$

$$\% \text{ O}_2 = \frac{1.26}{21.42} \times 100$$

$$= 5.9\%$$

$$\% \text{ CO}_2 = \frac{1.68}{21.42} \times 100$$

$$= 7.8\%$$

$$\% \text{ H}_2\text{O} = \frac{2.52}{21.42} \times 100$$

$$= 11.8\%$$

EXERCISE

Where required, take the composition by volume of O$_2$ in air as 21%, while that of N$_2$ as 79%. Also, the composition by mass of O$_2$ in air is 23.2%, while that of N$_2$ is 76.8%

1. 60kg of methane is combusted with 1100kg of air to give 85kg of carbon (IV) oxide and 42kg of carbon (II) oxide. Determine the percent excess air used.

(The molecular mass of air is 29kg/kmol)

2. During a combustion process 250 litres of ethene is burnt with 4800 litres of air to give 440 litres of carbon (IV) oxide and 120 litres of carbon (II) oxide. Calculate the percent excess air supplied.

3. A gas has the following composition by volume:
7% of CO_2, 15% of CO, 2.4% of C_2H_2, 16.5% of C_3H_8, 48% of H_2, 0.8% of O_2, and 10.3% of N_2. The gas is burnt to produce a stack gas containing the following composition by volume on a dry basis: 8.5% of CO_2, 6.9% of O_2, and 84.6% of N_2. Calculate the required and actual air supplied.

4. A fuel which contains carbon and hydrogen only, was burnt to produce a dry flue gas with the following composition by volume: 13.2% of CO_2, 7.1% of O_2 and 79.7% of N_2. Calculate:

(a). The composition of the fuel

(b). The percent excess air supplied

5. A petroleum gas was analyzed to contain the following composition by volume: 80% of C_2H_6, 15.2% of C_4H_{10}, and 4.8% of CO_2. The gas was burnt with 32% excess air. 90% of the hydrocarbons is converted to CO_2 while 10% is converted to CO. Calculate the composition of the flue gas.

6. A crude oil was analyzed to contain the following composition by mass: 84% of C, 12.4% of H_2 and 3.6% of S. If the stack gas is produced at 220°C and a pressure of 1.08×10^5 N/m², calculate:

(a). The composition of the stack gas from combustion of 100kg of the crude oil

(b). The volume of the stack gas.

(C = 12, H = 1, S = 32)

7. Propane is burnt with 10% excess air. Fractional conversion is 78%. Calculate the percentage composition of each of the components of the flue gas.

CHAPTER 7
BALANCES INVOLVING LIMITING REACTANTS

A limiting reactant is the component of a chemical reaction which is available for complete conversion in the reaction. It limits the excess reactant from reacting completely in the reaction.

An excess reactant is a reactant which is not completely used up in a reaction.

The percentage of the limiting reactant which actually reacts in a reaction is known as the degree of completion of the reaction.

Examples

1. Iron (II) sulphide was produced by heating five parts of iron with four parts of sulphur to produce 70% of iron (II) sulphide. The converter has a capacity of 1000kg. Determine:

(a). the limiting reactant

(b). the percent excess reactant

(c). the degree of completion of the reaction

Solution

(a). Let us use a basis of 1000kg of reactants.

The equation for the reaction is:

$$Fe + S \longrightarrow FeS$$

Five parts of Fe and four parts of S means that the ratio of Fe to S is 5 : 4

The total ration = 5 + 4 = 9

Therefore mass of Fe in feed = $\frac{5}{9}$ x 1000

$= 555.6 kg$

The mass of S in feed = $\frac{4}{9}$ x 1000

$= 444.4 kg$

Assuming that: Input = Output, then:

70% of FeS formed in the product means that:

$$\text{Mass of FeS in product} = \frac{70}{100} \times 1000$$

$$= 700\text{kg}$$

Converting these masses to kmols gives:

$$\text{kmols of Fe} = \frac{555.6}{56} \quad \text{(Note that the molecular mass of Fe = 56)}$$

$$= 9.9\text{kmols}$$

$$\text{kmols of S} = \frac{444.4}{32} \quad \text{(Note that the relative molecular mass of S = 32)}$$

$$= 13.9\text{kmols}$$

$$\text{kmols of FeS} = \frac{700}{56+32} = \frac{700}{88}$$

$$= 8.0\text{kmols}$$

The reaction is:

$$\text{Fe} + \text{S} \longrightarrow \text{FeS}$$

The reaction shows that 1kmol of Fe reacts with 1kmol of S to produce 1kmol of FeS

Therefore, 8kmols of FeS will be produced from:

$$8\text{Fe} + 8\text{S} \longrightarrow 8\text{FeS}$$

8kmols of FeS will be produced from 8kmols of Fe and 8kmols of S.

However, 9.9kmols of Fe supplied would also require 9.9kmols of S and not 13.9kmols of S. Hence S is present in excess.

This means that Fe is present in limited amount. Hence it is the limiting reactant.

(b). The excess reactant is S. 9.9kmols of S is required. But 13.9kmols is supplied. Therefore excess amount of S supplied = 13.9 - 9.9 = 4kmols

$$\text{Therefore, \% excess reactant} = \frac{\text{excess kmols}}{\text{required kmols}} \times 100$$

$$= \frac{4}{9.9} \times 100$$

= 40.4%

(c). Fe is the limiting reactant. The amount of Fe supplied is 9.9kmols.

The amount that reacted is 8kmols

Therefore, the degree of completion of the reaction is = $\frac{8}{9.9}$ x 100

= 80.8%

2. Calcium oxide is produced by the burning of calcium in oxygen. If 100kmols of calcium is mixed with 100kmols of oxygen, determine:

(a). the limiting reactant

(b). the percent excess of the excess reactant

(c). the composition of the product stream based on a complete combustion

(d). The composition of the product stream based on a 70% fractional conversion of the limiting reactant.

Solution

(a). The reaction is as shown below:

$$2Ca + O_2 \longrightarrow 2CaO$$

The reaction shows that 2kmols of Ca reacts with 1kmol of O_2 to produce 2kmols of CaO.

Therefore, 100kmols of Ca will require 50kmols of O_2.

But, 100kmols of O_2 is available. This shows that O_2 is present in excess.

Hence Ca is the limiting reactant.

(b). Of the 100kmols of O_2, only 50kmols will react with Ca.

Therefore the excess O_2 = 100 - 50

= 50kmols

Percent excess of the excess reactant (i.e. O_2) = $\frac{excess}{required}$ x 100

$$= \frac{50}{50} \times 100$$

$$= 100\%$$

(c). All the 100kmols of Ca will react with 50kmols of O_2 to produce 100kmols of CaO. Therefore the product will contain 50kmols of O_2 (unreacted O_2) and 100kmols of CaO.

Therefore, total kmols of product = 50 + 100

$$= 150 \text{kmols}$$

% of O_2 in product $= \frac{50}{150} \times 100$

$$= 33.3\%$$

% of CaO in product $= \frac{100}{150} \times 100$

$$= 66.7\%$$

(d). The reaction in terms of kmols is as follows:

100Ca + 50O_2 ----------> 100CaO

A 70% fractional conversion will give:

0.7(100Ca + 50O_2 ----------> 100CaO)

Or,   70Ca + 35O_2 ----------> 70CaO

This means that:

Unreacted Ca = 100 - 70

$$= 30 \text{kmols}$$

Unreacted O_2 = 100 - 35

$$= 65 \text{kmols}$$

Therefore the product stream contains:

70kmols of CaO (i.e. product formed)

30kmols of Ca (unreacted Ca)

65kmols of O_2 (unreacted O_2)

Hence total kmols in product stream = 70 + 30 + 65

$$= 165 \text{ kmols}$$

Therefore, % CaO in product $= \dfrac{70}{165} \times 100$

$$= 42.4\%$$

% of Ca in product $= \dfrac{30}{165} \times 100$

$$= 18.2\%$$

% of O_2 in product $= \dfrac{65}{165} \times 100$

$$= 39.4\%$$

3. An ore consist of pure $MgSO_4$. It is fused with carbon. The composition of the fusion mass is:

    8.5% of $MgSO_4$

    76.4% of MgS

    15.1% of C

The equation for the reaction is:

$$MgSO_4 + 4C \longrightarrow MgS + 4CO$$

Determine:

(a). the excess reactant

(b). percent excess reactant

(c). degree of completion of the reaction

<u>Solution</u>

(a). The molecular mass of $MgSO_4$ = 24 + 32 + (16 × 4)

$$= 24 + 32 + 64$$
$$= 120$$

The molecular mass of MgS = 24 + 32
$$= 56$$

The molecular mass of CO = 12 + 16
$$= 28$$

Therefore the equation can now be expressed along with the molecular mass as follows:

$$MgSO_4 + 4C \longrightarrow MgS + 4CO$$
$$120 + (4 \times 12) \longrightarrow 56 + (4 \times 28)$$
Or, $\quad 120 + 48 \longrightarrow 56 + 112$

This can be written as

$$120 kgMgSO_4 + 48 kgC \longrightarrow 56 kgMgS + 112 kgCO$$

Using MgS as the major product, 1 kg of MgS will be produced by dividing each substance by 56 (i.e. the molecular mass of MgS) as follows:

$$\frac{120}{56}kgMgSO_4 + \frac{48}{56}kgC \longrightarrow \frac{56}{56}kgMgS + \frac{112}{56}kgCO$$

Or, 1kgMgS is obtained from: $\frac{120}{56}kgMgSO_4 + \frac{48}{56}kgC$

Working on a basis of 100kg of the product, 76.4kg of MgS (i.e. composition from the question) is obtained from the reactants as follows:

$$76.4kgMgS \text{ is obtained from: } 76.4(\frac{120}{56}kgMgSO_4 + \frac{48}{56}kgC)$$

$$= 163.7kgMgSO_4 + 65.5kgC \quad \text{(obtained after simplifying the bracket above)}$$

This shows that in order to obtain 76.4kg of MgS, the reactants will be:

163.7kg of MgSO$_4$ and 65.5kg of C.

Since the product composition shows that 8.5kg of MgSO$_4$ is present in the product, then the total mass of MgSO$_4$ supplied = Reacted MgSO$_4$ + MgSO$_4$ in product (i.e. unreacted MgSO$_4$)

Therefore, MgSO$_4$ supplied = 163.7 + 8.5 = 172.2kg

Similarly, C supplied = 65.5 + 15.1

$$= 80.6 kg$$

Expressing these masses in kmols gives:

$$\text{kmols of MgSO}_4 \text{ supplied} = \frac{172.2}{120}$$

$$= 1.44 \text{kmols}$$

$$\text{kmols of C supplied} = \frac{80.6}{12}$$

$$= 6.72 \text{kmols}$$

Recall that the reaction is given by:

$$MgSO_4 + 4C \longrightarrow MgS + 4CO$$

This shows that 1kmol of MgSO$_4$ requires 4kmols of C

Therefore, 1.44kmols of MgSO$_4$ will require: $\frac{1.44}{1} \times 4$

$$= 5.76 \text{kmols of C}$$

However, the C supplied is 6.72kmols. This is above the required value of 5.76kmols. Hence, C is in excess.

Therefore the excess reactant is C.

(b). Kmols C supplied = 6.72kmols

Kmols C required = 5.76kmols

Therefore, % excess reactant = $\dfrac{\text{supplied} - \text{required}}{\text{required}} \times 100$

$$= \frac{6.72 - 5.76}{5.76} \times 100$$

$$= 16.7\%$$

(c). MgSO$_4$ is the limiting reactant. The amount of MgSO$_4$ supplied is 172.2kg. The amount that reacted is 163.7kg

Therefore, the degree of completion of the reaction is $= \dfrac{\text{Amount of MgSO}_4 \text{ reacted}}{\text{Amount of MgSO}_4 \text{ supplied}} \times 100$

$= \dfrac{163.7}{172.2} \times 100$

$= 95.1\%$

4. Iron (II) chloride was produced by heating 3 parts of iron with 2 parts of chlorine to produce 62% of iron (II) chloride. The converter has a capacity of 500kg. Determine:

(a). the limiting reactant

(b). the percent excess reactant

(c). the degree of completion of the reaction

Solution

(a). Let us use a basis of 500kg of reactants.

The equation for the reaction is:

$\qquad$ Fe + Cl$_2$ ----------> FeCl$_2$

3 parts of Fe and 2 parts of Cl$_2$ means that the ratio of Fe to Cl$_2$ is 3 : 2

The total ration = 3 + 2 = 5

Therefore mass of Fe in feed $= \dfrac{3}{5} \times 500$

$\qquad\qquad\qquad\qquad\qquad = 300$kg

The mass of Cl$_2$ in feed $= \dfrac{2}{5} \times 500$

$\qquad\qquad\qquad\qquad\qquad = 200$kg

Assuming that: Input = Output, then:

$\qquad$ 62% of FeCl$_2$ produced means that:

Mass of FeCl$_2$ produced $= \dfrac{62}{100} \times 500$

$\qquad\qquad\qquad\qquad\qquad = 310$kg

Converting these masses to kmols gives:

$$\text{kmols of Fe} = \frac{300}{56} \quad \text{(Note that the molecular mass of Fe = 56)}$$

$$= 5.4 \text{kmols}$$

$$\text{kmols of Cl}_2 = \frac{200}{71} \quad \text{(Note that the relative molecular mass of Cl}_2 = 2 \times 35.5 = 71\text{)}$$

$$= 2.8 \text{kmols}$$

$$\text{kmols of FeCl}_2 = \frac{310}{56+71}$$

$$= \frac{310}{127}$$

$$= 2.4 \text{kmols}$$

The reaction is:

$$\text{Fe} + \text{Cl}_2 \text{----------> FeCl}_2$$

The reaction shows that 1kmol of Fe reacts with 1kmol of Cl_2 to produce 1kmol of $FeCl_2$

Therefore, 2.4kmols of $FeCl_2$ will be produced from:

$$2.4\text{Fe} + 2.4\text{Cl}_2 \text{----------> } 2.4\text{FeCl}_2$$

2.4kmols of $FeCl_2$ will be produced from 2.4kmols of Fe and 2.4kmols of Cl_2.

However, 5.4kmols of Fe supplied would also require 5.4kmols of Cl_2 and not 2.8kmols of Cl_2. Hence Fe is present in excess.

This means that Cl_2 is present in limited amount. Hence, it is the limiting reactant.

(b). The excess reactant is Fe. 2.4kmols of Fe is required. But 5.4kmols is supplied. Therefore excess amount of Fe supplied = 5.4 - 2.4 = 3kmols

$$\text{Therefore, \% excess reactant} = \frac{\text{excess kmols}}{\text{required kmols}} \times 100$$

$$= \frac{3}{2.4} \times 100$$

$$= 125\%$$

(c). Cl_2 is the limiting reactant. The amount of Cl_2 supplied is 2.8kmols. The amount that reacted is 2.4kmols.

Therefore, the degree of completion of the reaction is = $\dfrac{\text{Amount of } Cl_2 \text{ reacted}}{\text{Amount of } Cl_2 \text{ supplied}}$ x 100

$= \dfrac{2.4}{2.8}$ x 100

= 85.7%

5. A tin stone consists of pure SnO_2. It is fused with charcoal. The composition of the fusion mass is:

    6.8% of SnO_2

    82% of Sn

    11.2% of C

The equation for the reaction is:

    $SnO_2 + 2C \longrightarrow Sn + 2CO$

Determine:

(a). the excess reactant

(b). percent excess reactant

(c). degree of completion of the reaction

Solution

(a). The molecular mass of SnO_2 = 119 + (16 x 2)   (Note that the atomic mass of Sn = 119)

                          = 119 + 32

                          = 151

The atomic mass of Sn = 119

The molecular mass of CO = 12 + 16

                        = 28

Therefore the equation can now be expressed along with the molecular mass as follows:

$$SnO_2 + 2C \longrightarrow Sn + 2CO$$
$$151 + (2 \times 12) \longrightarrow 119 + (2 \times 28)$$
Or, $151 + 24 \longrightarrow 119 + 56$

This can be written as

$$151 kg SnO_2 + 24 kg C \longrightarrow 119 kg Sn + 56 kg CO$$

Using Sn as the major product, 1kg of Sn will be produced by dividing each substance by 119 (i.e. the mass of Sn) as follows:

$$\frac{151}{119} kgSnO_2 + \frac{24}{119} kgC \longrightarrow \frac{119}{119} kgSn + \frac{56}{119} kgCO$$

Or, 1kgSn is obtained from: $\frac{151}{119} kgSnO_2 + \frac{24}{119} kgC$

Working on a basis of 100kg of the product, 82kg of Sn (i.e. composition from the question) is obtained from the reactants as follows:

$$82 kgSn \text{ is obtained from: } 82(\frac{151}{119} kgSnO_2 + \frac{24}{119} kgC)$$

$= 104 kgSnO_2 + 16.5 kgC$ (obtained after simplifying the bracket above)

This shows that in order to obtain 82kg of Sn, the reactants will be:

104kg of SnO_2 and 16.5kg of C.

Since the product composition shows that 6.8kg of SnO_2 is present in the product, then the total mass of SnO_2 supplied = Reacted SnO_2 + SnO_2 in product (i.e. unreacted SnO_2)

Therefore, SnO_2 supplied = 104 + 6.8 = 110.8kg

Similarly, C supplied = 16.5 + 11.2

$= 27.7$kg

Expressing these masses in kmols gives:

$$\text{kmols of } SnO_2 \text{ supplied} = \frac{110.8}{151}$$

$= 0.73$ kmols

$$\text{kmols of C supplied} = \frac{27.7}{12}$$

$$= 2.3 \text{ kmols}$$

Recall that the reaction is given by:

$$SnO_2 + 2C \longrightarrow Sn + 2CO$$

This shows that 1kmol of SnO_2 requires 2kmols of C

Therefore, 0.73kmols of SnO_2 will require: $\frac{0.73}{1} \times 2$

$$= 1.46 \text{ kmols of C}$$

However, the C supplied is 2.3kmols. This is above the required value of 1.46kmols. Hence, C is in excess.

Therefore the excess reactant is C.

(b). Kmols C supplied = 2.3kmols

Kmols C required = 1.46kmols

Therefore, % excess reactant = $\frac{\text{supplied} - \text{required}}{\text{required}}] \times 100$

$$= \frac{2.3 - 1.46}{1.46} \times 100$$

$$= 57.5\%$$

(c). SnO_2 is the limiting reactant. The amount of SnO_2 supplied is 110.8kg. The amount that reacted is 104kg

Therefore, the degree of completion of the reaction is = (Amount of SnO_2 reacted/Amount of SnO_2 supplied) $\frac{\text{Amount of SnO}_2 \text{ that reacted}}{\text{Amount of SnO}_2 \text{ supplied}} \times 100$

$$= \frac{104}{110.8} \times 100$$

$$= 93.9\%$$

EXERCISE

1. Iron (II) sulphide was produced by heating seven parts of iron with five parts of sulphur to produce 82% of iron (II) sulphide. The converter has a capacity of 2000kg. Determine:

(a). the limiting reactant

(b). the percent excess reactant

(c). the degree of completion of the reaction

(Fe = 56, S = 32)

2. Calcium oxide is produced by the burning of calcium in oxygen. If 500kmols of calcium is mixed with 200kmols of oxygen, determine:

(a). the limiting reactant

(b). the percent excess of the excess reactant

(c). the composition of the product stream based on a complete combustion

(d). The composition of the product stream based on a 90% fractional conversion of the limiting reactant.

3. An ore consist of pure $MgSO_4$. It is fused with carbon. The composition of the fusion mass is:

   7.8% of $MgSO_4$

   80% of MgS

   12.2% of C

The equation for the reaction is:

$$MgSO_4 + 4C \longrightarrow MgS + 4CO$$

Determine:

(a). the excess reactant

(b). percent excess reactant

(c). degree of completion of the reaction

4. Iron (III) oxide was produced by heating 6 parts of iron with 5 parts of oxygen to produce 75% of iron (III) oxide. The converter has a capacity of 2000kg. Determine:

(a). the limiting reactant

(b). the percent excess reactant

(c). the degree of completion of the reaction

5. A tin stone consists of pure SnO_2. It is fused with charcoal. The composition of the fusion mass is:

    6.6% of SnO_2

    85% of Sn

    8.4% of C

The equation for the reaction is:

$$SnO_2 + C \longrightarrow Sn + CO_2$$

Determine:

(a). the excess reactant

(b). percent excess reactant

(c). degree of completion of the reaction

CHAPTER 8
BALANCES ON SEPARATION PROCESSES

Material balances can be carried out on separation processes such as distillation processes, crystallization and filtration processes.

Examples

1. A distillation column is used to separate 1000kg/h of a mixture containing 60% water and 40% ethanol. The product at the top of the column contains 85% ethanol, while the product at the bottom contains 90% water. The vapour enters the condenser from the top of the column at 640kg/h and is returned to column as reflux. The rest is withdrawn. Calculate:

(a). the rate of formation of the top product and the bottom product

(b). the flow rate of obtaining ethanol in the top product

(c). the rate of production of ethanol in the bottom product

(d). the ratio of the amount refluxed to the product withdrawn

Solution

(a). Let us use a basis of 1000kg/h of the feed.

Let, F = feed rate = 1000kg/h

D = distillate rate (top product)

W = bottom product rate

x_F = mass fraction of ethanol in feed = 0.4

x_D = mass fraction of ethanol in the top product = 0.85

x_W = mass fraction of ethanol in the bottom product = 1 - 0.9 = 0.1 (water = 0.9)

The overall mass component balance is given by:

F = D + W   (since input = output)

1000 = D + W ....equation 1

The ethanol component balance is given by:

$$Fx_F = Dx_D + Wx_W$$

$$1000(0.4) = D(0.85) + W(0.1)$$

$$400 = 0.85D + 0.1W \quad \text{...........Equation 2}$$

From equation 1: $D = 1000 - W$ ....Equation 3

Substitute 1000 - W for D in equation 2

$$400 = 0.85D + 0.1W \quad \text{...........Equation 2}$$

$$400 = 0.85(1000 - W) + 0.1W$$

$$400 = 850 - 0.85W + 0.1W$$

$$400 = 850 - 0.75W$$

$$0.75W = 850 - 400$$

$$0.75W = 450$$

$$W = \frac{450}{0.75}$$

$$W = 600$$

From equation 3:

$$D = 1000 - W$$

$$= 1000 - 600$$

$$D = 400$$

Therefore the top product rate is 400kg/h, while the bottom product rate is 600kg/h

(b) The flow rate of obtaining ethanol in the top product = Dx_D

$$= 400 \times 0.85$$

$$= 340 \text{kg/h}$$

(c). The rate of producing ethanol in the bottom product = Wx_W

$$= 600 \times 0.1$$

$$= 60 \text{kg/h}$$

(d). The balance about the condenser is given by:

$$V = R + D$$

where V = vapour rate to the column, R = reflux rate, D = distillate rate (top product)

Therefore, V = R + D

$$640 = R + 400$$

$$R = 640 - 400$$

$$R = 240$$

Therefore the ratio of the amount refluxed to the product withdrawn which is the reflux ratio is given by:

Reflux ratio = $\dfrac{R}{D}$

$$= \dfrac{240}{400}$$

$$= 0.6$$

2. An alcohol-water mixture contains 50% by mass of alcohol. A distillation column is used to separate the mixture by a continuous process. If the composition of water in the top product is 15%, and it is 80% at the bottom, determine:

(a). the amount of alcohol in the top product

(b). what percentage of the feed is the top product

(c). the amount of alcohol in the top product as a percentage of alcohol in the feed

(d). the amount of water in the bottom product.

Solution

(a). The overall component balance on a basis of 100kg/h feed rate is given by:

$$F = D + W$$

$$100 = D + W \quad\text{.......................Equation 1}$$

The alcohol component balance is given by:

$$Fx_F = Dx_D + Wx_W$$

But $x_F = 0.5$, $x_D = 1 - 0.15 = 0.85$, and $x_W = 1 - 0.8 = 0.2$

Therefore, $Fx_F = Dx_D + Wx_W$

$$100(0.5) = D(0.85) + W(0.2)$$

$$50 = 0.85D + 0.2W \quad \text{.......................Equation 2}$$

From equation 1: $D = 100 - W$ ....Equation 3

Substitute $100 - W$ for D in equation 2

$$50 = 0.85D + 0.2W \quad \text{.......................Equation 2}$$

$$50 = 0.85(100 - W) + 0.2W$$

$$50 = 85 - 0.85W + 0.2W$$

$$0.85W - 0.2W = 85 - 50$$

$$0.65W = 35$$

$$W = 35/0.65$$

$$W = 53.8$$

From equation 3, $D = 100 - W$

$$= 100 - 53.8$$

$$D = 46.2$$

Top product, $D = 46.2$ kg/h, and bottom product, $W = 53.8$ kg/h

Amount of alcohol in the top product is $= Dx_D$

$$= 46.2 \times 0.85$$

$$= 39.3 \text{ kg/h}$$

(b). Feed = 100kg/h, top product = 46.2kg/h

Therefore percent of feed which is the top product $= \dfrac{46.2}{100} \times 100$

$$= 46.2\%$$

(c) Alcohol in feed = Fx_F

$$= 100 \times 0.5$$

$$= 50 \text{kg/h}.$$

Alcohol in the top product = 39.3kg/h

Therefore amount of alcohol in the top product as a percentage of alcohol in the feed is:

$$= \frac{39.3}{50} \times 100$$

$$= 78.6\%$$

(d) The amount of water in the bottom product = W x mass fraction of water in the bottom product = 53.8 x 0.8

$$= 43.0 \text{kg/h}$$

3. Water and air flow into a humidifier in which the water evaporates completely into the air. The entering air contains 1.5mole% water vapour. The outlet humidified air contains 20mole% water. Calculate the volume of water in m^3/min required to humidify 100mole/min of the entering air.

Solution

Let F_1 = moles/min of inlet water

F_2 = moles/min of inlet air

P = moles/min of product (humidified air)

Also, let m_1 = mole fraction of water in inlet water = 1 (since inlet water contains only water)

m_2 = mole fraction of water in inlet air = 0.015 (from 1.5% in the question)

m_3 = mole fraction of water in product = 0.2 (from 20%)

The overall component balance is given by:

$F_1 + F_2 = P$

$F_1 + 100 = P$ ....Equation 1

The water component balance is given by:

$$F_1 m_1 + F_2 m_2 = P m_3$$

$$F_1(1) + 100(0.015) = P(0.2)$$

$$F_1 + 1.5 = 0.2P \quad \text{...................Equation 2}$$

Substitute $F_1 + 100$ (equation 1) for P into equation 2. This gives:

$$F_1 + 1.5 = 0.2P \quad \text{...................Equation 2}$$

$$F_1 + 1.5 = 0.2(F1 + 100)$$

$$F_1 + 1.5 = 0.2F1 + 20$$

$$F_1 - 0.2F_1 = 20 - 1.5$$

$$0.8F_1 = 18.5$$

Therefore, $F_1 = \dfrac{18.5}{0.8}$

$$F_1 = 23.1 \text{ mole/min}$$

Recall that 1 mole of water = 18g (i.e. the molecular mass of water)

Therefore, 23.1 moles 0f water = 23.1 x 18

$$= 415.8g$$

Therefore mass of inlet water = 415.8g/min

Recall that: Density = $\dfrac{\text{mass}}{\text{volume}}$

$$1g/cm^3 = \dfrac{\text{mass in g}}{\text{volume in cm}^3}$$

Therefore volume in $cm^3 = \dfrac{\text{mass}}{1}$

$$= \dfrac{415.8}{1}$$

$$= 415.8 cm^3/min$$

Let us now convert this volume to m^3/min. In order to convert cm^3 to m^3, we divide by 100^3.

Therefore the volume in m³/min = $\frac{415.8}{100^3}$

$$= \frac{415.8}{10^6} \quad (100^3 = 100 \times 100 \times 100 = 10^6)$$

$$= 4.158 \times 10^{-4} \, m^3/min$$

4. An air conditioner takes in 250m³/min of air at 25°C, 720mmHg pressure and 92% relative humidity, and gives out the air at 12°C. Calculate:

(a). the flow rate of the outlet air

(b). the rate at which condensed water must be removed from the air conditioner.

(Take saturated vapour pressure of water at 25°C as 23.7mmHg, and at 12°C as 10.5mmHg)

Solutions

Let us calculate the mole fraction water in the inlet air. This can be obtained by using the formula:

$$y_1 = \frac{H_R}{100} \times \frac{P_t}{P_D}$$

where y_1 = mole fraction, H_R = relative humidity, P_t = saturated vapour pressure of water at air temperature, and P_D = saturated vapour pressure of water at dew point.

Therefore, $y_1 = \frac{H_R}{100} \times \frac{P_t}{P_D}$

$$= \frac{92}{100} \times \frac{23.7}{720}$$

$y_1 = 0.0303$

Therefore mole fraction of the dry air = 1 - 0.0303

$$= 0.9697$$

Let us also calculate the molar flow rate of the inlet air. In order to do this, recall the ideal gas equation: PV = mRT

Therefore, the standard condition can be expressed as:

$P_1V_1 = m_1 RT$ ....Equation 1

where P_1 = 760mmHg, V_1 = 22.4m³ (molar volume of gas), m_1 = 1kmol (since 1kmol = 22.4m³), and T_1 = 273k

The inlet condition of the air can be expressed as:

$$P_2V_2 = m_2RT \quad \text{......................Equation 2}$$

where P_2 = 720mmHg, V_2 = 250m³/min, T_2 = 273 + 25 = 298k, m_2 = ?

Dividing equation 1 by equation 2 gives:

$$\frac{P_1V_1}{P_2V_2} = \frac{m_1T_1}{m_2T_2} \quad \text{(R has cancelled out)}$$

This gives an expression relating initial and final conditions as follows:

$$\frac{P_1V_1m_2}{T_1} = \frac{P_2V_2m_1}{T_2}$$

Therefore, $m_2 = \dfrac{P_2V_2T_1m_1}{P_1V_1T_2}$

$$= \frac{720 \times 250 \times 1 \times 273}{760 \times 22.4 \times 298}$$

m_2 = 9.69kmols/min

At the outlet condition, the mole fraction of water in the air can be obtained by using Raoult's law as follows:

$$y_2P_D = P_t \quad (P_D = 720\text{mmHg}, P_t = 10.5)$$

Therefore, $y_2 = \dfrac{P_t}{P_D}$

$$= \frac{10.5}{720}$$

$$= 0.0146$$

Therefore mole fraction of dry outlet air = 1 - y_2

$$= 1 - 0.0146$$

$$= 0.9854$$

The dry air component balance is given by:

Input = Output

$(1 - y_1)m_2 = (1 - y_2)m_3$

where m_3 is the kmol/min of the outlet air.

Therefore, $0.9697 \times 9.69 = 0.9854 \times m_3$

Hence, $m_3 = \dfrac{0.9697 \times 9.69}{0.9854}$

$m_3 = 9.54$ kmols/min

Using this m_3, the outlet volume, V_3, can be calculated by using the expression below.

$$\dfrac{P_1 V_1 m_3}{T_1} = \dfrac{P_3 V_3 m_1}{T_3}$$

Therefore, $V_3 = \dfrac{P_1 V_1 m_3 T_3}{P_3 m_1 T_1}$

$= \dfrac{760 \times 22.4 \times 9.54 \times (273 + 12)}{720 \times 1 \times 273}$ (Note that $T_3 = 273 + 12$)

$= \dfrac{760 \times 22.4 \times 9.54 \times 285}{720 \times 273}$

$V_3 = 235.5 \text{m}^3/\text{min}$

(b). The water component balance is given by:

$m_2 y_1 = m_3 y_2 + m_4$ (where m_4 = kmol/min outlet water)

$9.69 \times 0.0303 = (9.54 \times 0.0146) + m_4$

$0.2936 = 0.1393 + m_4$

$m_4 = 0.2936 - 0.1393$

$m_4 = 0.154$

Therefore the condensed water must leave at a rate of 0.154kmols/min.

5. A gas containing vapour enters into a condenser. The partial pressure of the vapour in the gas is 800mmHg. The partial pressure of the vapour in the outlet gas is 780mmHg at a

temperature of 140°C. If the total pressure of the system is 860mmHg, calculate the volume of the gas leaving the condenser when 100kmols of the vapour condenses out.

Solution

From Raoult's law:

$$p = yP$$

Or, $y = \dfrac{p}{P}$

where p = partial pressure, and P = total pressure. At the inlet condition, the mole fraction of the vapour, y_1, is obtained as follows:

$$y_1 = \dfrac{p}{P}$$

$$= \dfrac{800}{860}$$

$$y_1 = 0.930$$

At the outlet condition, $y_2 = \dfrac{780}{860}$

$$y_2 = 0.907$$

The overall component balance is given by:

$$m_1 = m_2 + m_3$$

where m_1 = inlet gas, m_2 = outlet dry gas, and m_3 = outlet vapour = 100

Hence, $m_1 = m_2 + 100$ ....Equation 1

The vapour component balance is given by:

$$m_1 y_1 = m_2 y_2 + m_3 y_3 \quad (y_3 = 1, \text{ since only vapour is in } m_3)$$

Hence. $m_1 y_1 = m_2 y_2 + (100 \times 1)$

$$m_1 y_1 = m_2 y_2 + 100 \text{ ....Equation 2}$$

Substitute $m_2 + 100$ (in equation 1) for m_1 into equation 2. This gives:

$$m_1 y_1 = m_2 y_2 + 100 \text{ ....Equation 2}$$

$(m_2 + 100)0.930 = m_2(0.907) + 100$

$$0.93m_2 + 93 = 0.907m_2 + 100$$

$$0.93m_2 - 0.907m_2 = 100 - 93$$

$$0.023m_2 = 7$$

$$m_2 = \frac{7}{0.023}$$

$$m_2 = 304.3 \text{ kmols}$$

Therefore 304.3kmols of the dry gas leaves the condenser. This can be converted to volume by using the expression below.

$$\frac{P_1 V_1 m_2}{T_1} = \frac{P_2 V_2 m_1}{T_2}$$

where P_1 = 760mmHg, V_1 = 22.4m³, m_1 = 1kmol, T_1 = 273k, while the other quantities have their usual meanings.

Therefore, $V_2 = \dfrac{P_1 V_1 m_2 T_2}{P_2 m_1 T_1}$

$$= \frac{760 \times 22.4 \times 304.3 \times (273 + 140)}{860 \times 273}$$

$$= \frac{760 \times 22.4 \times 304.3 \times 413}{860 \times 273}$$

$$= 9113 \text{ m}^3$$

Therefore volume of the gas leaving the condenser is 9113m³

6. Wet paper pulp with a moisture content of 72% enters a continuous process dryer and leaves at a rate of 50kg/h with a moisture content of 8%. Dry air at 66°C and 76cmHg enters the dryer. The air and water vapour leaves the dryer at 55°C and 76cmHg. Calculate:

(a). the rate at which the wet pulp enters the dryer

(b). the flow rate of the inlet air in m³/h if the outlet air has a relative humidity of 45%

Solution

(a). Let us work with the rate of 50kg/h dried pulp. The dry pulp component balance is given by:

Input = output

$$m_2(1 - x_2) = (1 - x_4)m_4$$

where m_2 = inlet wet pulp flow rate in kg/h

m_4 = outlet pulp flow rate in kg/h = 50kg/h

x_2 = mass fraction of moisture in the inlet pulp = 0.72

x_4 = mass fraction of moisture in the outlet pulp = 0.08

Hence, $m_2(1 - x_2) = (1 - x_4)m_4$

$$m_2(1 - 0.72) = (1 - 0.08)m_4$$

$$0.28 m_2 = 0.92 \times 50$$

$$m_2 = \frac{46}{0.28}$$

$$m_2 = 164 \text{kg/h}$$

Therefore the wet pulp enters at 164kg/h

(b) Let y_3 be the mole fraction of water in the outlet air

Therefore, $y_3 P = \dfrac{H_R}{100} \times P_{H_2O}$ (at t = 55°C)

From tables, P_{H_2O} (at 55°C) = 119.6mmHg = 11.96cmHg

Therefore, $y_3 = \dfrac{H_R \times P_{H_2O}}{100 \times P}$

$$= \frac{45 \times 11.96}{100 \times 76}$$

$$y_3 = 0.0708$$

Let us now take water balance as follows:

Input = Output

Input water = Output in the dried pulp + Output water in air

Hence, $x_2 m_2 = x_4 m_4 + (y_3 m_3 \times$ molecular mass of water$)$

where m_3 is the outlet air flow rate in kmol/h. Note that y_3m_3 has been multiplied by the molecular mass of water in order to convert it from molar flow rate to mass flow rate since x_2m_2 and x_4m_4 are in mass flow rate.

$$x_2m_2 = x_4m_4 + (y_3m_3 \times \text{molecular mass of water})$$

$$0.72 \times 164 = (0.08 \times 50) + (0.0708 \times m_3 \times 18)$$

$$118.08 = 4 + 1.2744m_3$$

$$1.2744m_3 = 118.08 - 4$$

$$1.2744m_3 = 114.08$$

$$m_3 = \frac{114.08}{1.2744}$$

$$m_3 = 89.5 \text{kmol/h}$$

The dry air component balance is given by:

Input = Output

$$m_1 = (1 - y_3)m_3$$

where m_1 is the inlet air flow rate in kmols/h

Therefore, $m_1 = (1 - y_3)m_3$

$$m_1 = (1 - 0.0708) \times 89.5$$

$$= 0.9292 \times 89.5$$

$$m_1 = 83.2 \text{kmols/h}$$

Let us now convert this molar flow rate to volumetric flow rate as follows:

Recall that: $\dfrac{P_1V_1m_2}{T_1} = \dfrac{P_2V_2m_1}{T_2}$ (Refer to example 4)

Therefore, $V_2 = \dfrac{P_1V_1m_2T_2}{P_2m_1T_1}$

$$= \frac{76 \times 22.4 \times 83.2 \times (273 + 66)}{76 \times 1 \times 273}$$

$$= \frac{76 \times 22.4 \times 83.2 \times 339}{76 \times 1 \times 273}$$

$$= 2314 m^3/h$$

Therefore the flow rate of the inlet air is $2314 m^3/h$

Note that the inlet air conditions have been used as the final conditions (P_2, V_2, T_2, m_2), while the s.t.p conditions have been used as the initial conditions (P_1, V_1, T_1, m_1).

7. A solution of $MnSO_4.H_2O$ which is saturated at 72°C enters into a recycle stream. The recycled mixture is slightly evaporated and then cooled to 0°C. At 0°C, $MnSO_4.7H_2O$ crystals are formed, and this cooled stream is filtered. The product stream consists of 84% solid crystals and 16% solution. Calculate:

(a). the inlet flow rate that will produce 620kg/h of the crystals

(b). the rate of evaporation of the recycled mixture.

(Take the solubility of $MnSO_4.H_2O$ at 72°C as $61g/100gH_2O$, and the solubility of $MnSO_4.7H_2O$ at 0°C as $53g/100gH_2O$)

Solution

(a). Note that the feed is at 72°C, while the product is at 0°C. Let us convert the solubility to mass fraction.

Let x_F = mass fraction of $MnSO_4.H_2O$ in the feed

x_P = mass fraction of $MnSO_4.7H_2O$ in the product

Therefore, $x_F = \dfrac{61}{61+100}$

$= 0.3789$

Hence the mass fraction of water in feed = 1 - x_F

$= 1 - 0.3789$

$= 0.6211$

Similarly, $x_P = \dfrac{53}{53+100}$

$x_P = 0.3464$

Hence the mass fraction of water in the product = $1 - x_p$

$$= 1 - 0.3464$$
$$= 0.6536$$

Taking overall mass balance on a basis of 620kg/h of $MnSO_4.7H_2O$ produced gives:

Input = Output

$$F = 620 + P + W$$

where F = feed, P = 16% solution and W is the evaporated water

By applying simple proportion, P can be obtained as follows:

84% is equal to 620kg

Therefore, 16% will be equal to: $\frac{16}{84} \times 620 = 118kg$

Therefore, P = 118kg/h solution

Hence, F = 620 + 118 + W

$$F = 738 + W$$

Let us take $MnSO_4.H_2O$ component balance

Input = Output

$$Fx_F = 620\left(\frac{\text{molecular mass of } MnSO_4.H_2O}{\text{molecular mass of } MnSO_4.7H_2O}\right) + Px_P$$

Note that molecular mass of $MnSO_4.H_2O$ = 169, while the molecular mass of $MnSO_4.7H_2O$ = 277

Again, $Fx_F = 620\left(\frac{\text{molecular mass of } MnSO_4.H_2O}{\text{molecular mass of } MnSO_4.7H_2O}\right) + Px_P$

$$F(0.3789) = 620\left(\frac{169}{277}\right) + (118 \times 0.3464)$$

Note that $620\left(\frac{169}{277}\right)$ gives the mass of $MnSO_4.H_2O$ in the product crystals

$$0.3789F = 378.3 + 40.9$$

$$F = \frac{419.2}{0.3789}$$

$$F = 1106$$

Therefore the inlet feed flow rate is 1106kg/h

(b). Recall that: F = 738 + W

Therefore,   W = 1106 - 738

$$W = 368$$

Therefore the rate of evaporation of the recycled mixture is 368kgH_2O/h

EXERCISE

1. A distillation column is used to separate 2000kg/h of a mixture containing 68% water and 32% ethanol. The product at the top of the column contains 94% ethanol, while the product at the bottom contains 85% water. The vapour enters the condenser from the top of the column at 1250kg/h and is returned to column as reflux. The rest is withdrawn. Calculate:

(a). the rate of formation of the top product and the bottom product

(b). the flow rate of obtaining ethanol in the top product

(c). the rate of production of ethanol in the bottom product

(d). the ratio of the amount refluxed to the product withdrawn

2. An alcohol-water mixture contains 40% by mass of alcohol. A distillation column is used to separate the mixture by a continuous process. If the composition of water in the top product is 10%, and it is 88% at the bottom, determine:

(a). the amount of alcohol in the top product

(b). what percentage of the feed is the top product

(c). the amount of alcohol in the top product as a percentage of alcohol in the feed

(d) the amount of water in the bottom product.

3. Water and air flow into a humidifier in which the water evaporates completely into the air. The entering air contains 2.2mole% water vapour. The outlet humidified air contains 16mole% water. Calculate the volume of water in m^3/min required to humidify

500mole/min of the entering air.
(Take the molar mass of H_2O as 18g/mol and the density of H_2O as 1g/cm^3)

4. An air conditioner takes in 200m^3/min of air at 28°C, 740mmHg pressure and 85% relative humidity, and gives out the air at 8°C. Calculate:

(a). the flow rate of the outlet air

(b). the rate at which condensed water must be removed from the air conditioner.
(Take the SVP of water at 28°C as 28.3mmHg and at 8°C as 8mmHg)

5. A gas containing vapour enters into a condenser. The partial pressure of the vapour in the gas is 780mmHg. The partial pressure of the vapour in the outlet gas is 740mmHg at a temperature of 110°C. If the total pressure of the system is 850mmHg, calculate the volume of the gas leaving the condenser when 50kmols of the vapour condenses out.

6. Wet paper pulp with a moisture content of 46% enters a continuous process dryer and leaves at a rate of 200kg/h with a moisture content of 12%. Dry air at 62°C and 760mmHg enters the dryer. The air and water vapour leaves the dryer at 48°C and 760mmHg. Calculate:

(a). the rate at which the wet pulp enters the dryer

(b). the flow rate of the inlet air in m^3/h if the outlet air has a relative humidity of 41%
(Take the SVP of water at 48°C as 83.7mmHg)

7. A solution of $MnSO_4.H_2O$ which is saturated at 70°C enters into a recycle stream. The recycled mixture is slightly evaporated and then cooled to 0°C. At 0°C, $MnSO_4.7H_2O$ crystals are formed, and this cooled stream is filtered. The product stream consist of 92% solid crystals and 8% solution. Calculate:

(a). the inlet flow rate that will produce 800kg/h of the crystals

(b). the rate of evaporation of the recycled mixture.

(Take the solubility of $MnSO_4.H_2O$ at 70°C as 58g/100gH_2O, and the solubility of $MnSO_4.7H_2O$ at 0°C as 52g/100gH_2O)

CHAPTER 9
BALANCES ON SOLVENT EXTRACTION

In simple solvent extraction, a liquid is extracted (separated) from its mixture with another liquid in which both are miscible, i.e. they are soluble in each other. In order to carry out the extraction, a third liquid is used. The liquid to be extracted is soluble in this third liquid, while this third liquid is insoluble in the liquid that will be left unextracted. For example, acetone and hexane mix together to form a homogeneous mixture. When water is added to this mixture, the acetone which is soluble in water dissolves in the water, and hence it is extracted from its mixture with hexane. Water and hexane are insoluble (immiscible). Therefore, water has been used to extract acetone from its mixture with hexane. The water-acetone product is finally withdrawn.

The following examples cover balances on separation processes involving solvent extraction.

Examples

1. A 16% by weight acetone in water enters a single stage extraction unit at a feed rate of 1200litres/h. 92% of the acetone is to be extracted using chloroform at 25°C. Calculate the flow rate of the chloroform into the process.

(Take $K = \dfrac{(x_A)c \text{ phase}}{(x_A)w \text{ phase}} = 1.72$, where K is the distribution coefficient for the acetone - chloroform - water mixture. $(x_A)c$ is the mass fraction of acetone in chloroform, $(x_A)w$ is the mass fraction of acetone in water. Also, take density of acetone as 0.794g/cm^3 and the density of water as 1.0g/cm^3)

Solution

Let us take a basis of 1200litres/h of feed.

Therefore, the density of the feed is calculated as follows:

$$\frac{1}{\rho_F} = \frac{x_A}{\rho_A} + \frac{x_W}{\rho_W}$$

where ρ represents density, while F, A, and W, represent feed, acetone and water respectively. x_A is the mass fraction of acetone in feed, while x_W is the mass fraction of water in the feed.

Therefore, $\dfrac{1}{\rho_F} = \dfrac{x_A}{\rho_A} + \dfrac{x_W}{\rho_W}$

$= \dfrac{0.16}{0.794} + \dfrac{0.84}{1}$ (Note that $x_W = 100 - 16 = 84\% = 0.84$)

$= \dfrac{0.16 + 0.667}{0.794}$

$\dfrac{1}{\rho_F} = \dfrac{0.827}{0.794}$

Therefore $\rho_F = \dfrac{0.794}{0.827}$

Recall that: Density, $\rho = \dfrac{mass}{volume}$

Hence, mass = ρ x volume

Therefore, mass flow rate of feed = ρ_F x volumetric flow rate

$= \dfrac{0.794}{0.827} \times 1200$

$= 1152 kg/h$

Therefore, acetone in feed, $F_A = 0.16 \times 1152$

$= 184.3 kg/h$

Water in feed, $F_W = 0.84 \times 1152$

$= 967.7 kg/h$

Since 92% of the acetone is extracted from the feed, then:

$P_{AE} = 0.92 \times F_A$

$= 0.92 \times 184.3$

$= 169.6 kg/h$

where P_{AE} is the acetone extracted by chloroform.

Let us take the acetone component balance:

Input = Output

Hence, $F_A = P_{AL} + P_{AE}$

where P_{AL} is acetone left in the water.

Therefore, $184.3 = P_{AL} + 169.6$

$$P_{AL} = 184.3 - 169.6$$

$$P_{AL} = 14.7 \text{kg/h}$$

The mass fraction of acetone in the chloroform is given by:

$$(x_A)_C = \frac{P_{AE}}{P_{AE} + P_C}$$

where P_C is the chloroform in the product.

Hence, $(x_A)_C = \dfrac{169.6}{169.6 + P_C}$

$$= \frac{169.6}{169.6 + F_C}$$

where F_C is the chloroform feed rate. It is equal to P_C since it is assumed that chloroform is insoluble in water. Also, the mass fraction of acetone in water is given by:

$$(x_A)_W = \frac{P_{AL}}{P_{AL} + P_W}$$

where P_W is the water in the product. It is equal to F_W since all the water passes to the product.

Therefore, $(x_A)_W = \dfrac{P_{AL}}{P_{AL} + P_W}$

$$= \frac{14.7}{14.7 + F_W}$$

$$= \frac{14.7}{14.7 + 967.7}$$

$$= \frac{14.7}{982.4}$$

$(x_A)_W = 0.01496$

Recall that: $K = \dfrac{(x_A)c \text{ phase}}{(x_A)w \text{ phase}} = 1.72$

Therefore, $\dfrac{\frac{169.6}{169.6+F_C}}{0.01496} = 1.72$

$$\dfrac{169.6}{169.6+F_C} = 0.01496 \times 1.72$$

$$\dfrac{169.6}{169.6+F_C} = 0.02573$$

$$169.6 + F_C = \dfrac{169.6}{0.02573}$$

$$169.6 + F_C = 6591.5$$

$$F_C = 6591.5 - 169.6$$

$$F_C = 6422 \text{kg/h}$$

Therefore the flow rate of the chloroform into the process is 6422kg/h

2. A mixture of 72% by weight acetone and 28% by weight hexane is mixed with an equal mass of water. The overall mixture is shaken and allowed to stand. The acetone - water phase is withdrawn. The same amount of water is again added to the mixture left (i.e. hexane phase) and the process is carried out again. What percentage of the acetone in the feed is left unextracted in the hexane?

(Take K = $\dfrac{(x_A)_H}{(x_A)_W}$ = 0.34, where $(x_A)_H$ is the mass fraction of acetone in hexane, while $(x_A)_W$ is the mass fraction of acetone in water)

Solution

Here it is assumed that water and hexane are immiscible.

Let us take a basis of 100kg feed. This will consists of 72kg (72%) acetone and 28kg (28%) of hexane. The acetone component balance is given by:

$$F_1 = R_1 + E_1$$

where F = feed, R = raffinate, E = extract, and 1 represent stage 1. Note that the raffinate is the unextracted acetone in the hexane, while the extract is the extracted acetone in the water.

Therefore, $F_1 = R_1 + E_1$

$$72 = R_1 + E_1 \quad \text{......................Equation 1}$$

The mass fraction of acetone in hexane (i.e. the raffinate mass fraction) is given by:

$$(x_A)_H = \frac{R_1}{R_1 + 28} \quad \text{(Note that hexane = 28kg, and all the hexane remains in the raffinate)}$$

Similarly, the extract mass fraction is given by:

$$(x_A)_W = \frac{E_1}{E_1 + 100} \quad \text{(Note that water = 100kg, and all the water remains in the extract)}$$

Note that the mass of water is the same as the original feed mixture which is 100kg, since the feed was mixed with an equal mass of water.

Hence, $\quad \dfrac{(x_A)_H}{(x_A)_W} = 0.34 \quad$ (As given in the question)

$$\frac{\frac{R_1}{R_1 + 28}}{\frac{E_1}{E_1 + 100}} = 0.34 \quad \text{......................Equation 2}$$

From equation 1, $R_1 = 72 - E_1$ ....Equation 3

Substitute $72 - E_1$ for R_1 in equation 2. This gives:

$$\frac{\frac{R_1}{R_1 + 28}}{\frac{E_1}{E_1 + 100}} = 0.34 \quad \text{......................Equation 2}$$

$$\frac{\frac{72 - E_1}{72 - E_1 + 28}}{\frac{E_1}{E_1 + 100}} = 0.34$$

$$\frac{72 - E_1}{100 - E_1} \times \frac{E_1 + 100}{E_1} = 0.34$$

$0.34E_1(100 - E_1) = (72 - E_1)(E_1 + 100)$

$34E_1 - 0.34E_1^2 = 72E_1 + 7200 - E_1^2 - 100E_1$

$E_1^2 - 0.34E_1^2 + 34E_1 - 72E_1 + 100E_1 - 7200 = 0$

$0.66E_1^2 + 62E_1 - 7200 = 0 \quad$ (Quadratic equation)

Using the quadratic equation formula to solve this equation gives:

$$E_1 = \frac{-b \pm \sqrt{b^2 - 4ac}}{2a}$$

a = 0.66, b = 62, c = -7200

$$E_1 = \frac{-62 \pm \sqrt{62^2 - [4 \times 0.66 \times (-7200)]}}{2 \times 0.66}$$

$$= \frac{-62 \pm \sqrt{3844 + 19008}}{1.32}$$

$$= \frac{-62 \pm \sqrt{22852}}{1.32}$$

$$= \frac{-62 \pm 151}{1.32}$$

$$= \frac{89}{1.32} \quad \text{(The second answer is discarded since } E_1 \text{ cannot be negative)}$$

E_1 = 67.4kg

From equation 3:

$R_1 = 72 - E_1$

= 72 - 67.4

R_1 = 4.6kg

This raffinate, R_1 becomes the new feed for the second stage.

Taking acetone component balance in the second stage gives:

$R_1 = R_2 + E_2$

Hence, 4.6 = $R_2 + E_2$ ....Equation 4

In this second stage, the mass fraction of acetone in hexane phase is given by:

$(x_A)_H = \dfrac{R_2}{R_2 + 28}$ (All the hexane (28kg) will continue to be mixed with the raffinate)

The mass fraction of acetone in water phase is given by:

$(x_A)_W = \dfrac{E_2}{E_2 + 100}$ (Note that the same mass (100kg) of water was used in stage 2. All the water will always mix with the extracted acetone)

Therefore, similar to stage 1, the equilibrium in stage 2 is given by:

$$\frac{\frac{R_2}{R_2 + 28}}{\frac{E_2}{E_2 + 100}} = 0.34 \quad \text{......................Equation 5}$$

From equation 4, $R_2 = 4.6 - E_2$ ....Equation 6

Substitute $4.6 - E_2$ for R_2 in equation 5. This gives:

$$\frac{\frac{R_2}{R_2 + 28}}{\frac{E_2}{E_2 + 100}} = 0.34 \quad \text{......................Equation 5}$$

$$\frac{\frac{4.6 - E_2}{4.6 - E_2 + 28}}{\frac{E_2}{E_2 + 100}} = 0.34$$

$$\frac{4.6 - E_2}{32.6 - E_2} \times \frac{E_2 + 100}{E_2} = 0.34$$

$$0.34E_2(32.6 - E_2) = (4.6 - E_2)(E_2 + 100)$$

$$11.08E_2 - 0.34E_2^2 = 4.6E_2 + 460 - E_2^2 - 100E_2$$

$$E_2^2 - 0.34E_2^2 + 11.08E_2 + 100E_2 - 4.6E_2 - 460 = 0$$

$$0.66E_2^2 + 106.48E_2 - 460 = 0$$

Using quadratic equation formula to find E_2 gives:

$$E_2 = \frac{-b \pm \sqrt{b^2 - 4ac}}{2a}$$

$a = 0.66$, $b = 106.48$, $c = -460$

$$E_2 = \frac{-106.48 \pm \sqrt{106.48^2 - [4 \times 0.66 \times (-460)]}}{2 \times 0.66}$$

$$E_2 = \frac{-106.48 \pm \sqrt{11338 + 1214}}{1.32}$$

$$= \frac{-106.48 \pm \sqrt{12552}}{1.32}$$

$$= \frac{-106.48 \pm 112}{1.32}$$

$$= \frac{5.52}{1.32}$$ (The second answer is negative, so it cannot be our answer)

Therefore, E_2 = 4.2kg

From equation 6, R_2 = 4.6 - E_2

$$= 4.6 - 4.2$$

$$R_2 = 0.4\text{kg}$$

This is the unextracted acetone in the hexane.

% of acetone in the feed that is left unextracted in hexane is given by:

$$\frac{0.4}{72} \times 100 \quad \text{(Note that the acetone in feed = 72kg)}$$

$$= 0.56\%$$

3. A 24% by weight ethanol in benzene enters a single stage extraction unit at a feed rate of 500kg/h. 86% of the ethanol is to be extracted using water. Calculate the flow rate of the water into the process.

(Take $K = \frac{(x_E)_W \text{ phase}}{(x_E)_B \text{ phase}} = 1.84$, where $(x_E)_W$ is the mass fraction of ethanol in water, $(x_E)_B$ is the mass fraction of ethanol in benzene.

Solution

Ethanol in feed, F_E = 0.24 x 500

$$= 120\text{kg/h}$$

Benzene in feed, F_B = 0.76 x 500   (Note that 100 - 24 = 76% benzene in feed)

$$= 380\text{kg/h}$$

Since 86% of the ethanol is extracted from the feed, then:

P_{EE} = 0.86 x F_E

$$= 0.86 \times 120$$

$$= 103.2\text{kg/h}$$

where P_{EE} is the ethanol extracted by water.

Let us take the ethanol component balance:

Input = Output

Hence, $F_E = P_{EL} + P_{EE}$

where P_{EL} is ethanol left in the benzene.

Therefore, $120 = P_{EL} + 103.2$

$$P_{EL} = 120 - 103.2$$

$$P_{EL} = 16.8 \text{ kg/h}$$

The mass fraction of ethanol in the water is given by:

$$(x_E)_W = \frac{P_{EE}}{P_{EE} + P_W}$$

where P_W is the water in the product.

Hence, $(x_E)_W = \dfrac{103.2}{103.2 + P_W}$

$$= \frac{103.2}{103.2 + F_W}$$

where F_W is the water feed rate. It is equal to P_W since it is assumed that water is insoluble in benzene. Also, the mass fraction of ethanol in benzene is given by:

$$(x_E)_B = \frac{P_{EL}}{P_{EL} + P_B}$$

where P_B is the benzene in the product. It is equal to F_B since all the benzene passes to the product.

Therefore, $(x_E)_B = \dfrac{P_{EL}}{P_{EL} + P_B}$

$$= \frac{16.8}{16.8 + F_B}$$

$$= \frac{16.8}{16.8 + 380}$$

$$= \frac{16.8}{396.8}$$

$$(x_E)_B = 0.04234$$

Recall that: $K = \dfrac{(x_E)_W}{(x_E)_B} = 1.84$

Therefore, $\dfrac{\frac{103.2}{103.2 + F_W}}{0.04234} = 1.84$

$$\dfrac{103.2}{103.2 + F_W} = 0.04234 \times 1.84$$

$$\dfrac{103.2}{103.2 + F_W} = 0.07791$$

$$103.2 + F_W = \dfrac{103.2}{0.07791}$$

$$103.2 + F_W = 1324.6$$

$$F_W = 1324.6 - 103.2$$

$$F_W = 1221 \text{kg/h}$$

Therefore the flow rate of the water into the process is 1221kg/h

4. A 200kg mixture of 64% by weight acetone and 36% by weight pentane is mixed with 100kg of water. The overall mixture is shaken and allowed to stand. The acetone - water phase is withdrawn. 50kg of water is again added to the mixture left (i.e. pentane phase) and the process is carried out again. What is the total amount of acetone extracted?

(Take $K = \dfrac{(x_A)_p}{(x_A)_w} = 0.42$, where $(x_A)_p$ is the mass fraction of acetone in pentane, while $(x_A)_w$ is the mass fraction of acetone in water)

Solution

Here it is assumed that water and pentane are immiscible.

Mass of acetone in the feed = 0.64 x 200

= 128kg

Mass of pentane in the feed = 0.36 x 200

= 72kg

The acetone component balance is given by:

$$F_1 = R_1 + E_1$$

where F = feed, R = raffinate, E = extract, and 1 represent stage 1.

Therefore, $F_1 = R_1 + E_1$

$$128 = R_1 + E_1 \quad \text{......................Equation 1}$$

The mass fraction of acetone in pentane (i.e. the raffinate mass fraction) is given by:

$$(x_A)p = \frac{R_1}{R_1 + 72} \quad \text{(Pentane = 72kg. This is always mixed with the raffinate)}$$

Similarly, the extract mass fraction is given by:

$$(x_A)w = \frac{E_1}{E_1 + 100} \quad \text{(Water = 100kg. All the water is always mixed with the extract)}$$

Hence, $\dfrac{(x_A)p}{(x_A)w} = 0.42$ (As given in the question)

$$\frac{\frac{R_1}{R_1 + 72}}{\frac{E_1}{E_1 + 100}} = 0.42 \quad \text{......................Equation 2}$$

From equation 1, $R_1 = 128 - E_1$ ....Equation 3

Substitute $128 - E_1$ for R_1 in equation 2. This gives:

$$\frac{\frac{R_1}{R_1 + 72}}{\frac{E_1}{E_1 + 100}} = 0.42 \quad \text{......................Equation 2}$$

$$\frac{\frac{128 - E_1}{128 - E_1 + 72}}{\frac{E_1}{E_1 + 100}} = 0.42$$

$$\frac{128 - E_1}{200 - E_1} \times \frac{E_1 + 100}{E_1} = 0.42$$

$$0.42E_1(200 - E_1) = (128 - E_1)(E_1 + 100)$$

$$84E_1 - 0.42E_1^2 = 128E_1 + 12800 - E_1^2 - 100E_1$$

$$E_1^2 - 0.42E_1^2 + 84E_1 - 128E_1 + 100E_1 - 12800 = 0$$

$$0.58E_1^2 + 56E_1 - 12800 = 0 \quad \text{(Quadratic equation)}$$

Using the quadratic equation formula to solve this equation gives:

$$E_1 = \frac{-56 \pm \sqrt{56^2 - [4 \times 0.58 \times (-12800)]}}{2 \times 0.58}$$

$$= \frac{-56 \pm \sqrt{3136 + 29696}}{1.16}$$

$$= \frac{-56 \pm \sqrt{32832}}{1.16}$$

$$= \frac{-56 \pm 181.2}{1.16}$$

$$= \frac{125.2}{1.16} \quad \text{(The second answer is discarded since } E_1 \text{ cannot be negative)}$$

$E_1 = 108\,kg$

From equation 3:

$R_1 = 128 - E_1$

$= 128 - 108$

$R_1 = 20\,kg$

This raffinate, R_1, becomes the new feed for the second stage.

Taking acetone component balance in the second stage gives:

$R_1 = R_2 + E_2$

Hence, $20 = R_2 + E_2$ ....Equation 4

In this second stage, the mass fraction of acetone in pentane phase is given by:

$(x_A)p = \dfrac{R_2}{R_2 + 72}$ (All the pentane (72kg) will continue to be mixed with the raffinate)

The mass fraction of acetone in water phase is given by:

$(x_A)w = \dfrac{E_2}{E_2 + 50}$ (Note that the mass of water that was used in stage 2 is 50kg. All the water will always mix with the extracted acetone)

Therefore, similar to stage 1, the equilibrium in stage 2 is given by:

$$\frac{\frac{R_2}{R_2+72}}{\frac{E_2}{E_2+50}} = 0.42 \quad \text{...................Equation 5}$$

From equation 4, $R_2 = 20 - E_2$ ....Equation 6

Substitute $20 - E_2$ for R_2 in equation 5. This gives:

$$\frac{\frac{R_2}{R_2+72}}{\frac{E_2}{E_2+50}} = 0.42 \quad \text{.......................Equation 5}$$

$$\frac{\frac{20-E_2}{20-E_2+72}}{\frac{E_2}{E_2+50}} = 0.42$$

$$\frac{20-E_2}{92-E_2} \times \frac{E_2+50}{E_2} = 0.42$$

$$0.42E_2(92 - E_2) = (20 - E_2)(E_2 + 50)$$

$$38.64E_2 - 0.42E_2^2 = 20E_2 + 1000 - E_2^2 - 50E_2$$

$$E_2^2 - 0.42E_2^2 + 38.64E_2 + 50E_2 - 20E_2 - 1000 = 0$$

$$0.58E_2^2 + 68.64E_2 - 1000 = 0$$

Using quadratic equation formula to find E_2 gives:

$$E_2 = \frac{-68.64 \pm \sqrt{68.64^2 - [4 \times 0.58 \times (-1000)]}}{2 \times 0.58}$$

$$= \frac{-68.64 \pm \sqrt{4711 + 2320}}{1.16}$$

$$= \frac{-68.64 + 83.9}{1.16} \quad \text{(The second answer is negative, so it cannot be our answer)}$$

Therefore, $E_2 = 13.2$ kg

Hence the total amount of acetone extracted = $E_1 + E_2$

$$= 108 + 13.2$$

$$= 121.2 \text{kg}$$

EXERCISE

1. A 22% by weight acetone in water enters a single stage extraction unit at a feed rate of 1500litres/h. 95% of the acetone is to be extracted using chloroform at 25°C. Calculate the flow rate of the chloroform into the process.

(Take $K = \dfrac{(x_A)_C}{(x_A)_W} = 1.75$, where K is the distribution coefficient for the acetone -chloroform - water mixture. $(x_A)_C$ is the mass fraction of acetone in chloroform, $(x_A)_W$ is the mass fraction of acetone in water. Also, take density of acetone as $0.788g/cm^3$ and the density of water as $1.0g/cm^3$)

2. A mixture of 80% by weight acetone and 20% by weight hexane is mixed with an equal mass of water. The overall mixture is shaken and allowed to stand. The acetone - water phase is withdrawn. The same amount of water is again added to the mixture left (i.e. hexane phase) and the process is carried out again. What percentage of the acetone in the feed is left unextracted in the hexane.

(Take $K = \dfrac{(x_A)_H}{(x_A)_W} = 0.32$, where $(x_A)_H$ is the mass fraction of acetone in hexane, while $(x_A)_W$ is the mass fraction of acetone in water)

3. A 35% by weight ethanol in benzene enters a single stage extraction unit at a feed rate of 2000kg/h. 82% of the ethanol is to be extracted using water. Calculate the flow rate of the water into the process.

(Take $K = \dfrac{(x_E)_W}{(x_E)_B} = 1.68$, where $(x_E)_W$ is the mass fraction of ethanol in water, $(x_E)_B$ is the mass fraction of ethanol in benzene.

4. A 600kg mixture of 55% by weight acetone and 45% by weight pentane is mixed with 400kg of water. The overall mixture is shaken and allowed to stand. The acetone - water phase is withdrawn. 100kg of water is again added to the mixture left (i.e. pentane phase) and the process is carried out again. What is the total amount of acetone extracted.

(Take K = $\dfrac{(x_A)_P}{(x_A)_W}$ = 0.48, where $(x_A)_P$ is the mass fraction of acetone in pentane, while $(x_A)_W$ is the mass fraction of acetone in water)

5. A 1000kg mixture of 50% by weight acetone and 50% by weight pentane is mixed with 500kg of water. The overall mixture is shaken and allowed to stand. The acetone - water phase is withdrawn. 300kg of water is again added to the mixture left (i.e. pentane phase) and the process is carried out again. What is the total amount of acetone extracted?

(Take K = $\dfrac{(x_A)_P}{(x_A)_W}$ = 0.44, where $(x_A)_P$ is the mass fraction of acetone in pentane, while $(x_A)_W$ is the mass fraction of acetone in water)

CHAPTER 10
CALCULATIONS INVOLVING THE DETERMINATION OF FORMULA OF COMPOUNDS

The formula of a compound shows the number of atoms present in one molecule of the compound. This formula can either be empirical formula or molecular formula.

The empirical formula is a formula which shows the ratio of the number of atoms of elements present in a compound.

The molecular formula of a compound shows the exact number of atoms of element present in a compound. For example the molecular formula of benzene is C_6H_6, while its empirical formula is CH, which is a ratio of 1 : 1. Empirical formula can be determined from the constituent masses or percentage composition of the elements in the compound. Molecular formula can be obtained from empirical formula if the molecular mass of the compound is known. The examples given below show the various methods of finding the formulae of compounds.

Examples

1. 4.675kg of an oxide of copper on reduction yielded 4.15 of copper. Calculate the empirical formula of the oxide (Cu = 63.5, O = 16)

Solution

Mass of oxide = 4.675g

Mass of copper = 4.15g

Therefore, mass of oxygen = 4.675 - 4.15

= 0.525g

The calculation of the empirical formula can be set out as follows:

	Cu	O
Mass of element:	4.15	0.525
Moles of element:	$\frac{4.15}{63.5} = 0.0654$	$\frac{0.525}{16} = 0.0328$
Divide by the smallest moles	$\frac{0.0654}{0.0328} = 2$	$\frac{0.0328}{0.0328} = 1$

Therefore the ratio of the atoms is 2 : 1

Hence, the empirical formula is Cu_2O (This means 2 atoms of copper and 1 atom of oxygen)

2. On analysis, 1.24g of an oxide of sulphur was found to contain 0.62g of sulphur. Calculate the:

(a). percentage by mass of each element in the compound

(b). empirical formula of the compound

(S = 32, O = 16)

Solution

(a). % by mass of sulphur = $\frac{0.62}{1.24}$ x 100

= 50%

% by mass of oxygen = 100 - 50

= 50%

(b). The empirical formula is calculated as follows:

	S	O
% composition of element:	50%	50%
Ratio of atoms:	$\frac{50}{32}$ = 1.563	$\frac{50}{16}$ = 3.125
Divide each by the smallest ratio	$\frac{1.563}{1.563}$ = 1	$\frac{3.125}{1.563}$ = 2

Therefore, the empirical formula is SO_2. (This means 1 atom of sulphur and 2 atoms of oxygen)

3. The percentage composition of a compound is given by: Sodium 16.2%, Carbon 4.1%, Oxygen 16.9%, and water of crystallization 62.8%. Calculate the formula of the compound. (Na = 23, C = 12, O = 16, H = 1)

Solution

	Na	C	O	H₂O
% Composition	16.2%	4.1%	16.9%	62.8
Ratio of atoms	$\frac{16.2}{23} = 0.704$	$\frac{4.1}{12} = 0.342$	$\frac{16.9}{16} = 1.06$	$\frac{62.8}{18} = 3.49$
Divide by smallest ratio	$\frac{0.704}{0.342} = 2$	$\frac{0.342}{0.342} = 1$	$\frac{1.06}{0.342} = 3$	$\frac{3.49}{0.342} = 10$

This shows that the compound contains 2 atoms of Na, 1 atom of C, 3 atoms of O and 10 molecules of H_2O.

Hence the formula of the compound is $Na_2CO_3 \cdot 10H_2O$

4. 8cm³ of a hydrocarbon is put in a eudiometer tube and sparked with excess oxygen. After the combustion, the volume of the resulting gaseous mixture was found to be 102cm³. An introduction of a pellet of sodium hydroxide reduced the volume to 70cm³. After cooling the tube to room temperature, the volume of gas left was found to be 30cm³. Determine:

(a). the formula of the hydrocarbon

(b). the volume of oxygen sparked with the hydrocarbon

Solutions

From the question the following information can be obtained.

Volume of carbon (IV) oxide = volume of gas absorbed by sodium hydroxide = 102 - 70 = 32cm³

Volume of water = volume of gas condensed after cooling (Since the steam condensed to water) = 70 - 30 = 40cm³.

Note that a hydrocarbon contains carbon and hydrogen only, and it burns completely in oxygen to produce carbon (IV) oxide and steam (water).

Let the hydrocarbon be C_xH_y. Therefore the reaction can be written as follows.

$$C_xH_y + O_2 \longrightarrow xCO_2 + \frac{y}{2}H_2O$$ (Note that x and y have been used to balance the equation)

8cm³ ---------> 32cm³   40cm³

1Vol. ---------> 4Vol.   5Vol. (After dividing throughout by 8 to make C_xH_y 1 volume)

1 mole ----------> 4moles   5moles

Therefore, 1 mole C_XH_Y produces ----------> 4 moles CO_2 and 5 moles H_2O

This shows from the equation that x = 4 and $\frac{y}{2}$ = 5

If $\frac{y}{2}$ = 5, the y = 2 x 5

= 10

Therefore, x = 4 and y = 10.

Hence the hydrocarbon is C_XH_Y which is C_4H_{10} (when x is replaced by 4 and y is replaced by 10).

(b). The balanced equation using 1 mole of the hydrocarbon is given by:

C_4H_{10} + $\frac{13}{2}O_2$ -----> $4CO_2$ + $5H_2O$ (Use the total oxygen on the right to balance oxygen on the left)

Hence, 1mole + $\frac{13}{2}$moles ----> 4moles + 5 moles

Multiplying each of the value above by $8cm^3$ (i.e. volume of the hydrocarbon) gives:

$8cm^3$ + $52cm^3$ ----------> $32cm^3$ + $40cm^3$

This shows that the volume of oxygen that reacted is $52cm^3$. Recall that the final volume of gas left after cooling the mixture was $30cm^3$. This is the volume of the unreacted oxygen. Therefore the volume of oxygen sparked with the hydrocarbon is given by:

Reacted oxygen + unreacted oxygen

= 52 + 30

= $82cm^3$.

Therefore, $82cm^3$ of oxygen was sparked with the hydrocarbon.

5. $7.2cm^3$ of a hydrocarbon is sparked with $50.2cm^3$ of oxygen. After reaction and cooling of the vessel, the volume of the gaseous mixture was found to be $32.2cm^3$. Sodium hydroxide

was added to the reaction vessel. This reduced the gaseous mixture to 3.4cm^3. Determine the formula of the hydrocarbon.

Solution

Volume of hydrocarbon = 7.2cm^3

Volume of carbon (IV) oxide = 32.2 - 3.4

$$= 28.8 cm^3$$

Volume of unreacted oxygen (i.e. final gas left) = 3.4cm^3

Therefore, volume of reacted oxygen = 50.2 - 3.4

$$= 46.8 cm^3$$

Let the hydrocarbon be C_XH_Y. Therefore the reaction can now be written and balanced in terms of x and y as shown below.

$$C_XH_Y + O_2 \text{---------->} xCO_2 + \frac{y}{2}H_2O$$
$$7.2 + 46.8 \text{--------->} 28.8$$
$$1\text{mole} + 6.5\text{mole} \text{---->} 4\text{moles} \quad \text{(After dividing throughout by 7.2 to make } C_XH_Y \text{ 1 mole)}$$

The equation can now be written as follows:

$$C_XH_Y + 6\frac{1}{2}O_2 \text{---------->} 4CO_2 + \frac{y}{2}H_2O$$

Or, $C_XH_Y + \frac{13}{2}O_2 \text{---------->} 4CO_2 + 5H_2O$ (After balancing the oxygen atoms)

From the reaction above, there are 13 atoms ($\frac{13}{2}$ x 2 = 13) of oxygen on the left hand side. Therefore in order to balance the oxygen atoms (i.e. to also have 13 oxygen atoms on the right), there has to be 5 moles of H_2O on the right hand side. This is how we got the 5 in the reaction above.

Hence, $\frac{y}{2}H_2O = 5H_2O$

$$\frac{y}{2} = 5$$

Therefore, y = 2 x 5

y = 10

Similarly, $xCO_2 = 4CO_2$

Therefore, x = 4

Hence the hydrocarbon, C_XH_Y, is C_4H_{10}.

6. 20cm³ of a hydrocarbon was mixed with 110cm³ of oxygen and exploded. After cooling to room temperature, 80cm³ of gas was left. Addition of potassium hydroxide absorbed 40cm³ of the gas. Determine the formula of the hydrocarbon.

Solution

Volume of hydrocarbon = 20cm³

Volume of carbon (IV) oxide = 40cm³

Volume of unreacted oxygen = 80 - 40   (Note that unreacted oxygen is the gas left)

$\qquad\qquad$ = 40cm³

Therefore, volume of reacted oxygen = 110 - 40

$\qquad\qquad$ = 70cm³

The equation of the reaction is now written as follows:

$\qquad C_XH_Y + O_2 \text{----------> } xCO_2 + \frac{y}{2}H_2O$

$\qquad$ 20 $\quad$ 70 ----------> 40

$\qquad$ 1mole + $3\frac{1}{2}$moles --> 2moles $\qquad$ (After dividing each value by 20 to make C_XH_Y 1 mole)

Therefore, $C_XH_Y + \frac{7}{2}O_2 \text{----------> } 2CO_2 + 3H_2O$

From the reaction above, there are 7 atoms ($\frac{7}{2} \times 2 = 7$) of oxygen on the left hand side.
Therefore in order to balance the oxygen atoms (i.e. to also have 7 oxygen atoms on the right), there has to be 2 moles of CO_2 and 3 moles of H_2O on the right hand side. This is how we got the 2 and 3 in the reaction above.

Hence, by comparing the first and last equations, we see that: $\frac{y}{2} = 3$, which simplifies to y = 2 x 3 = 6.

Similarly x = 2.

Therefore the hydrocarbon, C_xH_y, is C_2H_6.

7. $12cm^3$ of a hydrocarbon is sparked with excess oxygen. After the combustion, the volume of the resulting gaseous mixture was found to be $55.2cm^3$. An introduction of a pellet of sodium hydroxide reduced the volume to $43.2cm^3$. After cooling the tube to room temperature, the volume of gas left was found to be $19.2cm^3$. Determine:

(a). the formula of the hydrocarbon

(b). the volume of oxygen sparked with the hydrocarbon

Solutions

From the question the following information can be obtained.

Volume of carbon (IV) oxide = volume of gas absorbed by sodium hydroxide = 55.2 - 43.2 = $12cm^3$

Volume of water = volume of gas condensed after cooling (Since the steam condensed to water) = 43.2 - 19.2 = $24cm^3$.

Let the hydrocarbon be C_xH_y. Therefore the reaction can be written as follows.

$C_xH_y + O_2$ ----------> $xCO_2 + \frac{y}{2}H_2O$ (Note that x and y have been used to balance the equation)

    $12cm^3$ ----------> $12cm^3$ $24cm^3$

    1Vol.   ----------> 1Vol.   2Vol. (After dividing throughout by 12 to make C_xH_y 1 volume)

    1 mole   ----------> 1moles 2moles

Therefore, 1 mole C_xH_y produces ----------> 1 mole CO_2 and 2 moles H_2O

This shows from the equation that x = 1 and $\frac{y}{2} = 2$

If $\frac{y}{2} = 2$, the y = 2 x 2

        = 4

Therefore, x = 1 and y = 4.

Hence the hydrocarbon, C_xH_y, is C_1H_4, which is CH_4

(b). The balanced equation using 1 mole of the hydrocarbon is given by:

$CH_4 + 2O_2$ ----------> $CO_2 + 2H_2O$ (Use the total oxygen on the right to balance oxygen on the left)

1mole + 2moles ----> 1mole + 2 moles

Multiplying each of the value above by 12 (i.e. volume of the hydrocarbon) gives:

$12cm^3 + 24cm^3$ ----------> $12cm^3 + 24cm^3$

This shows that the volume of oxygen that reacted is $24cm^3$. Recall that the final volume of gas left after cooling the mixture was $19.2cm^3$. This is the volume of the unreacted oxygen. Therefore the volume of oxygen sparked with the hydrocarbon is given by:

    Reacted oxygen + unreacted oxygen

= 24 + 19.2

= $43.2cm^3$.

Therefore, $43.2cm^3$ of oxygen was sparked with the hydrocarbon.

8. 1.56g of a hydrocarbon produced 5.28g of carbon (IV) oxide and 1.08g of water when burnt in air. Find:

(a). the empirical formula of the hydrocarbon

(b). its molecular formula if its molecular mass is 78

(C = 12, O = 16, H = 1)

Solution

(a). Molecular mass of CO_2 = 44, and atomic mass of carbon, C, is 12.

Therefore the mass of carbon in 5.28g of carbon (IV) oxide is given by simple proportion as follows:

$$\frac{12}{44} \times 5.24 = 1.44g$$

Similarly, the mass of hydrogen (H_2 = 2) in 1.08g of water (H_2O = 18) is given by:

$$\frac{2}{18} \times 1.08 = 0.12g$$

Therefore the empirical formula is calculated as follows:

	C	H
Mass of element:	1.44	0.12
Moles of element:	$\frac{1.44}{12} = 0.12$	$\frac{0.12}{1} = 0.12$
Divide each by the smallest mole:	$\frac{0.12}{0.12} = 1$	$\frac{0.12}{0.12} = 1$

Therefore the empirical formula is CH.

(b). The molecular formula is calculated as follows:

$(CH)n = 78$ (where n is a whole number)

$(12 + 1)n = 78$ (since C = 12 and H = 1)

$13n = 78$

$n = \frac{78}{13}$

$n = 6$

Hence, $(CH)n = (CH)_6 = C_6H_6$.

Therefore the molecular formula is C_6H_6.

9. A hydrocarbon contains 83.3% of carbon. Its density at s.t.p is 6.43g/dm^3. Find the:

(a). empirical formula of the hydrocarbon

(b). molecular formula of the hydrocarbon

Solution

(a). The percentage of hydrogen in the hydrocarbon is:

100 - 83.3 = 16.7%

The empirical formula is obtained as follows:

	C	H
% composition:	83.3	16.7
Ratio of atoms:	$\frac{83.3}{12} = 6.94$	$\frac{16.7}{1} = 16.7$
Divide by the smallest ratio:	$\frac{6.94}{6.94} = 1$	$\frac{16.7}{6.94} = 2.4$

This ratio 1 : 2.4 for the atoms, is not a whole number ratio. So we have to make it whole number. In order to make this ratio to be a whole number ratio, we look for the smallest number that can multiply 2.4 to give a whole number. The number is 5. So, multiply 1 and 2.4 by 5 to give:

(1 x 5) : (2.4 x 5)

= 5 : 12

Therefore the empirical formula is C_5H_{12}.

(b). A density of 6.44g/dm^3 means that 1dm^3 of the compound has a mass of 6.43g.

Therefore, at s.t.p: 22.4dm^3 of the compound will have a mass of:

22.4 x 6.43

= 144g

Hence, the molecular mass of the compound is 144

The molecular formula is calculated as follows:

(C_5H_{12})n = 144   (where n is a whole number)

[(12 x 5) + (1 x 12)]n = 78   (since C = 12 and H = 1)

(60 + 12)n = 144

72n = 144

$$n = \frac{144}{72}$$

$$n = 2$$

Hence, $(C_5H_{12})n = (C_5H_{12})_2 = C_{10}H_{24}$.

Therefore the molecular formula is $C_{10}H_{24}$.

10. 1.0g of an organic compound containing carbon, hydrogen and oxygen, on oxidation produced 1.38g of carbon (IV) oxide and 1.12g of water. 0.8g vapour of the compound occupied a volume of 0.56dm³ at s.t.p. Determine:

(a). the empirical formula of the compound

(b). its molecular formula

Solution

(a). Mass of carbon in 1.38g of carbon (IV) is given by:

$\frac{12}{44}$ x 1.38   (Note that C = 12 and CO_2 = 44)

= 0.376g

Similarly, the mass of hydrogen in 1.12g of water is given by:

$\frac{2}{18}$ x 1.12   (Note that H_2 = 2 and H_2O = 18)

= 0.124g

Therefore, the mass of oxygen in the compound is given by:

1 - (0.376 + 0.124)   (From the question sample of compound is 1g)

= 1 - 0.5

= 0.5g

We now calculate the empirical formula as follows:

C   H   O

| Mass of element: | 0.376 | 0.124 | 0.5 |

| Moles of element: | $\frac{0.376}{12} = 0.0313$ | $\frac{0.124}{1} = 0.124$ | $\frac{0.5}{16} = 0.0313$ |

| Divide by smallest mole: | $\frac{0.0313}{0.0313} = 1$ | $\frac{0.124}{0.0313} = 4$ | $\frac{0.0313}{0.0313} = 1$ |

Therefore, the empirical formula is CH_4O

(b). $0.56dm^3$ of the compound has a mass of 0.8g

Therefore, $22.4dm^3$ of the compound will have a mass of:

$\frac{22.4}{0.56} \times 0.8$ (This is by simple proportion)

= 32g

Therefore, the molecular mass of the compound is 32g

The molecular formula is calculated as follows:

$(CH_4O)n = 32$ (where n is a whole number)

$(12 + (1 \times 4) + 16)n = 32$ (since C = 12, H = 1, and O = 16)

$(12 + 4 + 16)n = 32$

$32n = 32$

$n = \frac{32}{32}$

$n = 1$

Hence, $(CH_4O)n = (CH_4O)_1 = CH_4O$

Therefore the molecular formula is CH_4O

EXERCISE

1. 2.34kg of an oxide of copper on reduction yielded 2.08kg of copper. Calculate the empirical formula of the oxide (Cu = 63.5, O = 16)

2. On analysis, 3.72g of an oxide of sulphur was found to contain 1.86g of sulphur. Calculate the:

(a). percentage by mass of each element in the compound

(b). empirical formula of the compound

(S = 32, O = 16)

3. The percentage composition of a compound is given by: Sodium 15.9%, Carbon 4.3%, Oxygen 16.7%, and water of crystallization 63.1%. Calculate the formula of the compound. (Na = 23, C = 12, O = 16, H = 1)

4. 10cm^3 of a hydrocarbon is put in a eudiometer tube and sparked with excess oxygen. After the combustion, the volume of the resulting gaseous mixture was found to be 92cm^3. An introduction of a pellet of sodium hydroxide reduced the volume to 52cm^3. After cooling the tube to room temperature, the volume of gas left was found to be 12cm^3. Determine:

(a). the formula of the hydrocarbon

(b). the volume of oxygen sparked with the hydrocarbon

5. 3.6cm^3 of a hydrocarbon is sparked with 75cm^3 of oxygen. After reaction and cooling of the vessel, the volume of the gaseous mixture was found to be 66cm^3. Sodium hydroxide was added to the reaction vessel. This reduced the gaseous mixture to 48cm^3. Determine the formula of the hydrocarbon.

6. 40cm^3 of a hydrocarbon was mixed with 250cm^3 of oxygen and exploded. After cooling to room temperature, 190cm^3 of gas was left. Addition of potassium hydroxide absorbed 120cm^3 of the gas. Determine the formula of the hydrocarbon.

7. 12cm^3 of a hydrocarbon is sparked with excess oxygen. After the combustion, the volume of the resulting gaseous mixture was found to be 76cm^3. An introduction of a pellet of sodium hydroxide reduced the volume to 52cm^3. After cooling the tube to room temperature, the volume of gas left was found to be 16cm^3. Determine:

(a). the formula of the hydrocarbon

(b). the volume of oxygen sparked with the hydrocarbon

8. 6.24g of a hydrocarbon produced 21.12g of carbon (IV) oxide and 4.32g of water when burnt in air. Find:

(a). the empirical formula of the hydrocarbon

(b). its molecular formula if its molecular mass is 26

(C = 12, O = 16, H = 1)

9. A hydrocarbon contains 83.3% of carbon. Its density at s.t.p is $3.22g/dm^3$. Find the:

(a). empirical formula of the hydrocarbon

(b). molecular formula of the hydrocarbon

10. 2g of an organic compound containing carbon, hydrogen and oxygen, on burning, produced 2.76g of carbon (IV) oxide and 2.24g of water. 1.6g vapour of the compound occupied a volume of $2.24dm^3$ at s.t.p. Determine:

(a). the empirical formula of the compound

(b). its molecular formula

CHAPTER 11
PRESSURE IN LIQUID

Pressure is defined as force per unit area. Its S.I unit is N/m² which is called Pascal (Pa). Other popular units of pressure are "mmHg" and "atm". Pressure is given by:

$$P = \frac{F}{A}$$

The pressure exerted by the air above us is called atmospheric pressure. Its value is $1.013 \times 10^5 N/m^2$, 760mmHg or 1atm.

In engineering calculations, absolute pressure is normally used. It is given by:

$$P_{absolute} = P_{atmospheric} + P_{gauge}$$

where P_{gauge} is the gauge pressure

In a liquid, the pressure at a level in the liquid is given by:

$$P = \rho g h$$

where ρ is the density of the liquid, h is the depth of level above the liquid surface, while g is the acceleration due to gravity.

Also recall that: density, $\rho = \frac{mass}{volume}$ or $\rho = \frac{m}{V}$

U - Tube Manometer

This is a device used for measuring gas pressure or a differential pressure. When the two limbs of the manometer are open, atmospheric pressure acts on them. If the difference in height between the liquid levels in both limbs is h, then the pressure being measured is a measure of the gauge pressure. However, if one limb of the tube is sealed, then the h will give a measure of the absolute pressure.

The sealed tube can function as a barometer when the pressure at the open limb is atmospheric.

Manometric Pressure Difference

When a manometer open at both ends is connected across an orifice plate in a pipeline through which a fluid of density ρ is flowing, then the pressure drop or pressure difference across the orifice plate is given by:

$P_1 - P_2 = (\rho_F - \rho)gh_F$

Or $\Delta P = (\rho_F - \rho)gh_F$

where P_1 is the upstream pressure before the orifice plate, P_2 is the downstream pressure after the orifice plate, ΔP is the pressure difference/pressure drop, ρ_F is the density of the liquid in the manometer, and h_F is the difference in height between the liquid levels in the two limbs of the manometer.

If the fluid whose pressure is being measured is a gas, then $(\rho_F - \rho)$ becomes approximately ρ_F since ρ (i.e. the gas density) is very small. Hence the equation above becomes:

$\Delta P = \rho_F g h_F$

The Inclined Manometer

For an inclined manometer, h_F in the equation of a vertical manometer has to be replaced by the expression:

$h_F = L\sin\theta$ (Since $\sin\theta = \dfrac{\text{opposite}}{\text{hypotenuse}} = \dfrac{h_F}{L}$, i.e. trigonometric ratio for an inclined plane)

where L is the slant height (hypotenuse) of the manometer, while θ is the angle of inclination of the manometer to the horizontal. Generally, $\sin\theta = 0.1$.

The manometric equation for an inclined manometer is given by:

$P_1 - P_2 = (\rho_F - \rho)gL\sin\theta$ (h_F is replaced by $L\sin\theta$)

For a gas pressure measurement, the equation becomes:

$P_1 - P_2 = \rho_F g L\sin\theta$

Examples

1. A piece of wood in the shape of a cylinder is 2.8m long, and it floats vertically in a water tank. If the part of the wood above the water surface is 1.5m,

(a). calculate the density of the wood.

(b). If the liquid in the tank is one whose density is unknown and the portion of the wood above the liquid surface is 1.9m, calculate the density of the liquid.

(Density of water = $1000 kg/m^3$)

Solution

(a). **Method 1:**

From Archimedes's principle, when a body floats in a liquid, the mass of the liquid displaced is equal to the mass of the body.

Let the cross-sectional area of the wood be A. Length of wood submerged in the water is:

    2.8 - 1.5 = 1.3m

Therefore, volume of water displaced by the wood is:

    V = A x 1.3   [Note that volume = cross-sectional area x length (or height)]

    V = 1.3A

Recall that: $\rho = \dfrac{m}{V}$

Hence, $m = \rho V$

Therefore mass of water displaced = 1000 x 1.3A

                                = 1300A

Similarly, mass of wood = $\rho_W \times V_W$ (where ρ_W is the density of wood and V_W is the volume of wood)

But, V_W = length of wood x cross sectional area of wood

        = 2.8A

Hence, mass of wood = $\rho_W \times V_W$

                         = ρ_W x 2.8A

                         = 2.8Aρ_W

      Mass of wood = mass of water displaced   (Archimedes's principle)

Therefore,   2.8Aρ_W = 1300A

$$\rho_W = \dfrac{1300A}{2.8A}$$

$$\rho_W = 464.3 \text{kg/m}^3$$

Method 2

From method 1 above, it can be established that the density of a solid of uniform cross-sectional area floating in a liquid is given by:

$$\text{Density of floating solid} = \frac{\text{Density of liquid} \times \text{submerged length of solid}}{\text{Total length of solid}}$$

Therefore, density of wood, $\rho_W = \dfrac{1000 \times 1.3}{2.8}$

$$\rho_W = 464.3 \text{kg/m}^3$$

(b). **Method 1**

$$\text{Density of floating solid} = \frac{\text{Density of liquid} \times \text{submerged length of solid}}{\text{Total length of solid}}$$

$$464.3 = \frac{\rho_L \times (2.8 - 1.9)}{2.8} \quad (\rho_L = \text{density of liquid})$$

$$464.3 = \frac{0.9 \rho_L}{2.8}$$

Therefore, $\rho_L = \dfrac{464.3 \times 2.8}{0.9}$

$$\rho_L = 1444.5 \text{kg/m}^3$$

Method 2

When a solid float in a liquid, the length of the solid submerged is inversely proportional to the density of the liquid. This is expressed as follows:

$$L \, \alpha \, \frac{1}{\rho}$$

When such a body floats in two different liquids, then the lengths submerged and the densities of the liquids are related by the expression below:

$$\frac{L_A}{L_B} = \frac{\rho_B}{\rho_A}$$

By using water and the new liquid in question (b) above, we have:

$$\frac{L_{Wt}}{L_{NL}} = \frac{\rho_{NL}}{\rho_{Wt}}$$

where NL represents new liquid, Wt represents water, ρ represents density, while L represents length of solid submerged.

Therefore, $\dfrac{L_{Wt}}{L_{NL}} = \dfrac{\rho_{NL}}{\rho_{Wt}}$

$$\dfrac{2.8 - 1.5}{2.8 - 1.9} = \dfrac{\rho_{NL}}{1000}$$

$$\dfrac{1.3}{0.9} = \dfrac{\rho_{NL}}{1000}$$

$$\rho_{NL} = \dfrac{1000 \times 1.3}{0.9}$$

$\rho_{NL} = 1444.4 \text{kg/m}^3$ (As obtained before)

2. A storage tank is used to supply water to a vat. A pressure of at least 2.2atm gauge pressure is needed at the inlet of the vat.

(a). Determine the minimum height from the water level in the tank to the inlet of the vat.

(b). If diesel of density 0.84kg/Litre is used, what will be the height required.

Solutions

(a). Recall that: $\Delta P = \rho_F g h_F$

But, $\Delta P = 2.2$ atm

Convert this pressure to N/m² by multiplying it by $1.013 \times 10^5 \text{N/m}^2$

Therefore, 2.2atm = $(2.2 \times 1.013 \times 10^5)$ N/m²

$= 2.229 \times 10^5$ N/m²

Therefore, $\Delta P = \rho_F g h_F$

$$h_F = \dfrac{\Delta P}{\rho_F g}$$

$$= \dfrac{2.229 \times 10^5}{1000 \times 9.8}$$ (Note that density of water = 1000kg/m³ and g = 9.8m/s²)

$h_F = 22.7$m

(b). Density of diesel = 0.84kg/Litre. Multiply it by 1000 to convert it to density in kg/m³.

Therefore, density of diesel = 0.84 x 1000

$$= 840 \text{kg/m}^3$$

Therefore, $\Delta P = \rho_F g h_F$

Hence, $h_F = \dfrac{\Delta P}{\rho_F g}$

$$= \dfrac{2.229 \times 10^5}{840 \times 9.8}$$

$h_F = 27.1\text{m}$

3. A manometer closed at one end contains liquid of unknown density. The difference between the liquid levels is 5.2m, when the barometer reading is 780mmHg.

(a). Determine the density of the liquid

(b). The same liquid is used in a manometer connected across an orifice plate in a pipeline through which water is flowing. If the difference in levels of the liquid in the manometer is 25cm with the same barometric reading, calculate the pressure drop in mmHg.

(c). What is the pressure in mmHg downstream of the pipe?

Solutions

(a). P_1 = 780mmHg. This can be converted to pressure in N/m² as follows:

$P_1 = (\dfrac{780}{760} \times 1.013 \times 10^5)$ (Since 760mmHg = 1.013 × 10⁵N/m²)

$P_1 = 1.04 \times 10^5 \text{N/m}^2$

P_2 = 0, since one end of the tube is closed.

Hence, $P_1 - P_2 = \rho_F g h_F$

$P_1 = \rho_F g h_F$ (Since $P_2 = 0$)

$\rho_F = \dfrac{P_1}{g h_F}$

$$= \frac{1.04 \times 10^5}{9.8 \times 5.2}$$

$$\rho_F = 2041 \text{kg/m}^3$$

Therefore the density of the liquid is 2041kg/m³

(b). $\Delta P = (\rho_F - \rho)gh_F$ ($h_F = \frac{25}{100} = 0.25$m)

$$= (2041 - 1000) \times 9.8 \times 0.25$$

$$= 2550 \text{N/m}^2$$

In order to convert this pressure to pressure in mmHg, divide it by 1.013 x 10⁵ and multiply the value by 760. This gives:

$$\frac{2550}{1.013 \times 10^5} \times 760$$

$$= 19.1 \text{mmHg}$$

(c) $\Delta P = P_1 - P_2$ (P_2 is the pressure downstream)

$$P_2 = P_1 - \Delta P$$

$$= 780 - 19.1$$

$$P_2 = 760.9 \text{mmHg}$$

Therefore the pressure downstream the pipe is 760.9mmHg

4. A manometer open at both ends contains three different liquids. The middle liquid y, forms a J shape in the tube such that the difference between its two levels is h_1. The top liquid z on the right arm of the tube makes a height of h_2. The liquid x at the left arm of the tube is on liquid y, and at the shorter arm of the J shape of y. The tops of x and z are at the same level.

(a). Find an equation for this manometer.

(b). Calculate the pressure at the left open arm of the tube given that the pressure on the right is 1.2 x 10⁵N/m² and h_1 = 30cm, h_2 = 40cm, ρ_x = 840kg/m³, ρ_y = 1350kg/m³, and ρ_z = 1000kg/m³.

Solutions

(a). $\Delta P = (\rho_F - \rho)gh_F$

Since there are three liquids in this manometer, $(\rho_F - \rho)gh_F$ will be computed into two separate values and then added.

With ρ_X as the reference density, (since it is the lowest), we compute $(\rho_F - \rho)gh_F$ for liquid y as follows:

$(\rho_Y - \rho_X)gh_1$

Similarly, $(\rho_F - \rho)gh_F$ is computed for liquid z as follows:

$(\rho_Z - \rho_X)gh_2$

Adding these two expressions gives the pressure drop and the equation for the manometer as follows:

$\Delta P = (\rho_Y - \rho_X)gh_1 + (\rho_Z - \rho_X)gh_2$

(b). $h_1 = 30cm = (\frac{30}{100})m = 0.3m$

$h_2 = 40cm = (\frac{40}{100})m = 0.4m$

$\Delta P = (\rho_Y - \rho_X)gh_1 + (\rho_Z - \rho_X)gh_2$

$P_1 - P_2 = (\rho_Y - \rho_X)gh_1 + (\rho_Z - \rho_X)gh_2$

$ = [(1350 - 840) \times 9.8 \times 0.3] + [(1000 - 840) \times 9.8 \times 0.4]$

$ = 1499.4 + 627.2$

$P_1 - P_2 = 2127$

$P_1 = 2127 + P_2$

$P_1 = 2127 + (1.2 \times 10^5)$

$P_1 = 122127 N/m^2$

5. An inclined manometer of length 15cm is inclined at an angle of $10°$. It is used in an orifice plate which is inserted in a pipeline through which a gas is flowing. Calculate the pressure difference across the orifice plate if the manometer contains mercury of density $13600 kg/m^3$.

Solution

For an inclined manometer we have:

$$\Delta P = (\rho_F - \rho)gL\sin\theta$$

$$\Delta P = \rho_F gL\sin\theta \quad \text{(Since } \rho \text{ is very small when compared to } \rho_F\text{)}$$

$$= 13600 \times 9.8 \times 0.15 \times \sin 10 \quad (15cm = 0.15m)$$

$$= 13600 \times 9.8 \times 0.15 \times 0.1736$$

$$\Delta P = 3471 N/m^2$$

This can be converted to pressure in mmHg as follows:

$$\Delta P = \frac{3471}{1.013 \times 10^5} \times 760$$

$$\Delta P = 26.0 mmHg.$$

EXERCISE

(Take the value of g as $9.8 m/s^2$)

1. A piece of wood in the shape of a cylinder is 1.6m long, and it floats vertically in a water tank. If the part of the wood above the water surface is 0.9m,

(a). calculate the density of the wood.

(b). If the liquid in the tank is one whose density is unknown and the portion of the wood above the liquid surface is 0.4m, calculate the density of the liquid.

(Density of water = $1000 kg/m^3$)

2. A storage tank is used to supply water to a vat. A pressure of at least 1.8atm gauge pressure is needed at the inlet of the vat.
(Density of water = $1000 kg/m^3$)

(a). Determine the minimum height from the water level in the tank to the inlet of the vat.

(b). If petrol of density 0.78kg/Litre is used, what will be the height required.

3. A manometer closed at one end contains liquid of unknown density. The difference between the liquid levels is 3.5m, when the barometer reading is 740mmHg.

(a). Determine the density of the liquid

(b). The same liquid is used in a manometer connected across an orifice plate in a pipeline through which water is flowing. If the difference in levels of the liquid in the manometer is 40cm with the same barometric reading, calculate the pressure drop in mmHg.

(c). What is the pressure in mmHg downstream of the pipe?

4. A manometer open at both ends contains three different liquids. The middle liquid q, forms a J shape in the tube such that the difference between its two levels is h_1. The top liquid r on the right arm of the tube makes a height of h_2. The liquid p at the left arm of the tube is on liquid q, and at the shorter arm of the J shape of q. The tops of p and r are at the same level.

(a). Find an equation for this manometer.

(b). Calculate the pressure at the left open arm of the tube given that the pressure on the right is $1.6 \times 10^5 N/m^2$ and $h_1 = 20cm$, $h_2 = 35cm$, $\rho_p = 790kg/m^3$, $\rho_q = 1200kg/m^3$, and $\rho_r = 950kg/m^3$.

5. An inclined manometer of length 24cm is inclined at an angle of 14°. It is used in an orifice plate which is inserted in a pipeline through which a gas is flowing. Calculate the pressure difference across the orifice plate if the manometer contains mercury of density $13600kg/m^3$.

CHAPTER 12
HUMIDITY AND WATER VAPOUR IN THE AIR

Humidity is the amount of water vapour present in the air. When air contains the maximum amount of water vapour that it can carry, then the air is said to be saturated at that air temperature and pressure.

At saturation, the expression for Raoult's law applies as follows:

$$p = yP$$

where p is the partial pressure of the saturated vapour, y is the mole fraction of the vapour, while P is the pressure of the moist air.

The expression above can also be expressed as:

$$yP = P_{H_2O}(T)$$

where $P_{H_2O}(T)$ is the vapour pressure of water at the temperature, T, while P is the pressure of the moist air.

Relative Humidity

This is the ratio of the amount of water in the air to the amount of water that will saturate the air at a particular temperature and pressure. It is usually expressed as a percentage. It is expressed mathematically as follows:

$$H_R = \frac{P_{H_2O}}{P_{satd}} \times 100$$

where H_R = relative humidity, P_{H_2O} = partial pressure of water vapour in the air, and P_{satd} = pressure of the saturated air at the prevailing temperature.

Molal Humidity

This is the ratio of the number of moles of water vapour to the number of moles of the vapour free air. It is expressed as:

$$H_m = \frac{\text{Number of moles of water vapour}}{\text{Number of moles of dry air}}$$

Or, $$H_m = \frac{\text{Mole fraction of water vapour}}{\text{Mole fraction of dry air}}$$

It can also be expressed as:

$$H_m = \frac{P_{H_2O}}{P - P_{H_2O}}$$

where P = total pressure, while P_{H_2O} = partial vapour pressure.

Absolute Humidity

This is the amount of water vapour present in a specific amount of air. It is expressed as grams of water vapour per cubic metre of air, i.e. g/m^3. It can also be expressed as grams of water vapour per kilogram of air, i.e. g/kg. Mathematically, it is expressed as:

$$H_A = \frac{\text{Mass of water vapour}}{\text{Mass of dry air}})$$

Examples

1. Water and air are contacted in a closed pot at a temperature of 70°C and a pressure of 760mmHg. Calculate the molar composition of the air-vapour mixture.

Solution

From steam tables, the saturated vapour pressure of water at 70°C is 236.8mmHg.

Therefore, $yP = P_{H_2O}(T)$

$$y = \frac{P_{H_2O}(70°C)}{P}$$

$$= \frac{236.8}{760}$$

Hence, $y_{H_2O} = 0.3116$

Therefore, the mole fraction of dry air is:

$y_{Air} = 1 - y_{H_2O}$

$= 1 - 0.3116$

$y_{Air} = 0.6884$

2. On a particular day, the temperature of the environment was found to be 28°C at a pressure of 730mmHg and relative humidity of 90%. Calculate:

(a). the mole fraction of water vapour in the air

(b). the molal humidity

(c). the dew point

(d). the absolute humidity

Solutions

(a). Raoult's law indicates that:

$$yP = P_{H_2O} \quad \text{...................Equation 1}$$

But relative humidity is given by:

$$H_R = \frac{P_{H_2O}}{P_{H_2O}(t°C)} \times 100$$

Therefore, $P_{H_2O} = \frac{H_R P_{H_2O}(t°C)}{100}$ ....Equation 2

Note that P_{H_2O} = partial pressure of water vapour, while $P_{H_2O}(t°C)$ = saturated vapour pressure at t°C.

Comparing equations 1 and 2 above, shows that they are both equal to P_{H_2O}. Therefore:

$$yP = \frac{H_R P_{H_2O}(t°C)}{100}$$

Hence, $y = \frac{H_R P_{H_2O}(t°C)}{100 \times P}$ ....Equation 3

From steam tables, P_{H_2O}(at 28°C) = 28.7mmHg. Substituting known values into equation 3 above gives:

$$y = \frac{H_R P_{H_2O}(t°C)}{100 \times P} \quad \text{....................Equation 3}$$

$$y = \frac{90 \times 28.7}{100 \times 730}$$

y_{H_2O} = 0.0354

(b). Molal humidity is given by:

H_m = mole fraction of water vapour/mole fraction of dry air

$$= \frac{0.0354}{1 - 0.0354}$$

$$= \frac{0.0354}{0.9646}$$

= 0.0367mol H₂O/mol dry air

(c). The dew point is the temperature at which this 0.0354 moles of water will saturate the air.

$$P_{H_2O}(t°C)_{\text{Dew Point}} = y_{H_2O} P$$

$$= 0.0354 \times 730$$

$$P_{H_2O}(t°C)_{\text{Dew Point}} = 25.8$$

From steam tables, the temperature that corresponds to this saturated vapour pressure is 26.2.

Therefore, the dew point, $t_{\text{Dew Point}}$ = 26.2°C.

(d). The absolute humidity is given by:

$$H_A = \frac{\text{mass of water vapour}}{\text{mass of dry air}}$$

Recall that: Number of moles = $\frac{\text{mass}}{\text{molar mass}}$

Therefore, mass of water vapour = Number of moles x molar mass of water

= 0.0354 x 18   (Note that molecular mass of H₂O = 18)

= 0.6372

Similarly, mass of dry air = moles of dry air x molar mass of air

= (1 - 0.0354) x 29   (Note that the molecular mass of air = 29)

= 0.9646 x 29

= 27.9734

Therefore, $H_A = \frac{\text{mass of water vapour}}{\text{mass of dry air}}$

$$= \frac{0.6372}{27.9734}$$

H_A = 0.0228g H₂O/g dry air

3. Air of relative humidity 90% enters into a condenser at a temperature of 65°C and a pressure of 750mmHg, and leaves the condenser at a temperature of 5°C. What is the volume of water that condenses out in 1 hour?

Solution

Basis: 1m³/min air at 65°C and 750mmHg

Volume of inlet air = 1m³/min, (i.e. the volumetric flow rate)

Temperature of inlet air = 65°C

Pressure of inlet air = 750 mmHg

Relative humidity of inlet air = 90%

Temperature of outlet air = 5°C

Also, let:

y_1 = mole fraction of water in inlet air

m_1 = molar flow rate of inlet air in kmol/min

y_2 = mole fraction of water in outlet air

m_2 = molar flow rate of outlet air in kmol/min

m_3 = molar flow rate of condensed water out of the unit in kmol/min

Let us calculate y_1.

Recall that: $y_1 = \dfrac{H_R \times P_{H_2O} \text{ (at 65°C)}}{100 \times P}$ (As explained in example 2 above)

From tables it can be obtained that:

P_{H_2O} at 65°C = 190 mmHg

Substituting into the equation above gives:

$$y_1 = \dfrac{90 \times 190}{100 \times 750}$$

$y_1 = 0.228$

Therefore, mole fraction of the dry air in the inlet air = 1 - y_1

$1 - 0.228 = 0.772$

Let us calculate the molar flow rate of the inlet air. Recall the ideal gas equation:

$PV = mRT$ ……………Equation 1

At standard conditions, P = 760mmHg, V = 22.4m³, (molar volume of gas), T = 273K, m = 1kmol (because 1kmol = 22.4m³)

At the inlet condition we have:

$P_1V_1 = m_1RT_1$ ………………Equation (2)

Where P_1 = 750mmHg, V_1 = 1m³/min, T_1 = 273 + 65 = 338K, m_1 = ?

Dividing equation 1 by equation 2 gives:

$$\frac{PV}{P_1V_1} = \frac{mT}{m_1T_1}$$ (Note that R has cancelled out)

Therefore, $m_1 = \frac{P_1V_1mT}{PVT_1}$

$$= \frac{750 \times 1 \times 1 \times 273}{760 \times 22.4 \times 338}$$

m_1 = 0.0356 kmol/min

At the outlet conditions, y_2 can be obtained from Raoult's law as follows:

$y_2P = P_{H_2O}$ (5°C)

Therefore, $y_2 = \frac{P_{H_2O}}{P}$

From tables, P_{H_2O} (at 5°C) = 6.63mmHg

Therefore, $y_2 = \frac{6.63}{750}$

y_2 = 0.00884

Therefore mole fraction of the dry air in the outlet air is:

$1 - y_2 = 1 - 0.00884 = 0.9912$

Taking dry air balance:

Input dry air = Output dry air

$(1-y_1)m_1 = (1 - y_2)m_2$

$0.772 \times 0.0356 = 0.9912 m_2$

$m_2 = \dfrac{0.772 \times 0.0356}{0.9912}$

$m_2 = 0.0277$ kmol/min

Taking water balance

Input water = Output water

$m_1 y_1 = m_2 y_2 + m_3$ (m_3 = molar flow rate of condensed water out of the unit in kmol/min)

∴ $m_3 = m_1 y_1 - m_2 y_2$

   $= (0.0356 \times 0.228) - (0.0277 \times 0.00884)$

$m_3 = 0.00787$ kmol/min

This water obtained can be converted to mass as follows:

0.00787 kmol $= (0.00787 \times 1000)$ mols

$= 7.87$ mols

But, Number of moles $= \dfrac{mass}{Molar\ mass}$

$7.87 = \dfrac{mass}{18}$ (Since $H_2O = 18$ g/mol)

∴ Mass $= 7.87 \times 18$

$= 141.7$ g

Therefore, 141.7g/min of water will condense out of the condense

Water that condensed out in 1 hour

Since in 1 minute, 141.7g of water is condensed out,

Therefore, in 1 hour, water that will condense out is given by:

141.7 x 60   (1 hour = 60 minutes)

= 8502g

= 8.502kg

8.502kg of water will condense in 1 hour

This mass can also be converted to litres to give:

8.502 Litres.   (Because the density of water = 1kg/L)

Therefore, 8.502 Litres of water will condense out of the unit in 1 hour

4. Air of relative humidity 50% enters into a condenser at a temperature of 35°C and a pressure of 680mmHg, and leaves the condenser at a temperature of 5°C.

(a). What is the mass of water that condenses out of the unit in 1min?

(b). What is the volume (in litres) of water that condenses out of the unit in 1 hour?

Solution

Basis: 1m³/min air at 35°C and 680mmHg

Volume of inlet air = 1m³/min, (i.e. the volumetric flow rate)

Temperature of inlet air = 35°C

Pressure of inlet air = 680 mmHg

Relative humidity of inlet air be = 50%

Temperature of outlet air = 5°C

Also, let:

y_1 = mole fraction of water in inlet air

m_1 = molar flow rate of inlet air in kmol/min

y_2 = mole fraction of water in outlet air

m_2 = molar flow rate of outlet air in kmol/min

m_3 = molar flow rate of condensed water out of the unit in kmol/min

Let us calculate y_1.

Therefore, $y_1 = \dfrac{H_R \times P_{H_2O}(\text{at } 35°C)}{100 \times P}$

From steam tables P_{H_2O} at 35°C = 42.8 mmHg

Substituting into the equation above gives:

$y_1 = \dfrac{50 \times 42.8}{100 \times 680}$

$y_1 = 0.0315$

Therefore, mole fraction of the dry air in the inlet air = $1 - y_1$

$= 1 - 0.0315 = 0.9685$

Let us calculate the molar flow rate of the inlet air. Recall the ideal gas equation:

$PV = mRT$ ....Equation 1

At standard conditions, P = 760mmHg, V = 22.4m³, (molar volume of gas), T = 273K, m = 1kmol (because 1kmol = 22.4m³)

At the inlet condition we have:

$P_1V_1 = m_1RT_1$ ....Equation 2

Where P_1 = 680mmHg, V_1 = 1m³/min, T_1 = 273 + 35 = 308K, m_1 = ?

Dividing equation 1 by equation 2 gives:

$\dfrac{PV}{P_1V_1} = \dfrac{mT}{m_1T_1}$ (Note that R has cancelled out)

Therefore, $m_1 = \dfrac{P_1V_1mT}{PVT_1}$

$= \dfrac{680 \times 1 \times 1 \times 273}{760 \times 22.4 \times 308}$

m_1 = 0.0354 kmol/min

At the outlet conditions, y_2 can be obtained from Raoult's law as follows:

$$y_2 P = P_{H_2O}$$

Therefore, $y_2 = \dfrac{P_{H_2O}}{P}$

From tables, P_{H_2O} (at 5°C) = 6.63mmHg

Therefore, $y_2 = \dfrac{6.63}{680}$

$$y_2 = 0.00975$$

Therefore mole fraction of the dry air in the outlet air is:

$$1 - y_2 = 1 - 0.00975$$

$$= 0.9903$$

<u>Taking dry air balance:</u>

Input dry air = Output dry air

$$(1-y_1)m_1 = (1 - y_2)m_2$$

$$0.9685 \times 0.0354 = 0.9903 m_2$$

$$m_2 = \dfrac{0.9685 \times 0.0354}{0.9903}$$

$$m_2 = 0.0346 \text{ kmol/min}$$

<u>Taking water balance</u>

Input water = Output water

$$m_1 y_1 = m_2 y_2 + m_3$$

∴ $m_3 = m_1 y_1 - m_2 y_2$

$$= (0.0354 \times 0.0315) - (0.0346 \times 0.00975)$$

$$m_3 = 0.000778 \text{ kmol/min}$$

This water obtained can be converted to mass as follows:

$$0.000778 \text{kmol} = (0.000778 \times 1000) \text{mols}$$

$$= 0.778 \text{ mols/min}$$

But, Number of moles $= \dfrac{\text{mass}}{\text{Molar mass}}$

$$0.778 = \dfrac{\text{mass}}{18} \quad \text{(Since } H_2O = 18\text{g/mol)}$$

∴ Mass = 0.778 x 18

$$= 14.0\text{g/min}$$

Therefore, 14.0g of water will condense out of the unit in 1 minute.

(b). Water that condensed out in 1 hour can be obtained as follows.

Since in 1 minute, 14.0g of water is condensed,

Therefore, in 1 hour, water that will condense is given by:

14 x 60   (1 hour = 60 minutes)

= 840g

= 0.84kg

0.84kg of water will condense in 1 hour

This mass can be converted to litres to give:

0.84 Litres.   (Because the density of water = 1kg/L)

Therefore, 0.84 Litres of water will condense out of the unit in 1 hour.

EXERCISE

1. Water and air are contacted in a closed vessel at a temperature of 68°C and a pressure of 760mmHg. Calculate the molar composition of the air-vapour mixture.
(SVP of water at 68°C = 214.2mmHg)

2. On a particular day, the temperature of the environment was found to be 34°C at a pressure of 700mmHg and relative humidity of 60%. Calculate:

(a). the mole fraction of water vapour in the air

(b). the molal humidity

(c). the dew point

(d). the absolute humidity

(SVP of water at 34°C = 39.9mmHg)

3. Air of relative humidity 60% enters into a condenser at a temperature of 45°C and a pressure of 770mmHg, and leaves the condenser at a temperature of 5°C. What is the volume of water that condenses out in 30 minutes?
(SVP of water at 45°C = 71.9mmHg and at 5°C = 6.5mmHg)

4. Air of relative humidity 80% enters into a condenser at a temperature of 40°C and a pressure of 760mmHg, and leaves the condenser at a temperature of 10°C.

(a). What is the mass of water that condenses out of the unit in 30min?

(b). What is the volume (in litres) of water that condenses out of the unit in 1 day?
(SVP of water at 40°C = 55.3mmHg and at 10°C = 9.2mmHg)

5. On a particular day, the temperature of the environment was found to be 22°C at a pressure of 740mmHg and relative humidity of 35%. Calculate:

(a). the mole fraction of water vapour in the air

(b). the molal humidity

(c). the dew point

(d). the absolute humidity

(SVP of water at 22°C = 19.8mmHg)

CHAPTER 13
EQUILIBRIUM REACTION CALCULATIONS

A reaction is said to be in chemical equilibrium when the rate of the forward reaction is equal to the rate of the backward reaction.

Laws of Chemical Equilibrium

1. If a reversible reaction is represented as $mA + nB \rightleftharpoons pC + qD$, where A and B are the reactants, C and D are the products, while m, n, p, q, are their number of moles respectively, then the equilibrium constant, K_C, is given by:

$$K_C = \frac{[C^p][D^q]}{[A^m][B^n]}$$

K_C is the equilibrium constant in terms of molar concentration, because A, B, C and D in the equation above are concentrations in mol/dm^3 or mol/litre.

2. In the case of gaseous reactions, concentrations are usually expressed in partial pressure. In this case, the equilibrium constant is written as K_P. Let us consider the reaction below.

$$2NO_{(g)} + O_{2(g)} \rightleftharpoons 2NO_{2(g)}$$

The equilibrium constant, K_P, in terms of partial pressure is given by:

$$K_P = \frac{(P_{NO_2})^2}{(P_{NO})^2 (P_{O_2})}$$

while the equilibrium constant, K_C, in terms of concentration is given by:

$$K_C = \frac{[NO_2]^2}{[NO]^2[O_2]}$$

For most reactions the values of K_C and K_P are different.

3. Equilibrium constant K, is also related to Gibbs free energy, G. The Gibbs free energy change, ΔG, for a reaction at a particular temperature is given by:

$\Delta G = -RT \ln K$

where R is the molar gas constant, T is the temperature in Kelvin, while K is the equilibrium constant.

Free Gibbs energy change can also be expressed in terms of change in enthalpy, ΔH, and change in entropy, ΔS, as follows:

$$\Delta G = \Delta H - T\Delta S$$

where T is the temperature in Kelvin.

The equilibrium constant, in terms of Gibbs free energy change is given by:

$$K_{eq} = e^{-\Delta G/RT}$$ (This is obtained by making K the subject of the formula $\Delta G = -RT\ln K$)

4. Consider the reaction: $2SO_2 + O_2 \rightleftharpoons 2SO_3$.

Its equilibrium constant is given by:

$$K_1 = \frac{[SO_3]^2}{[SO_2]^2[O_2]}$$

When the reaction above is multiplied by ½, the reaction becomes:

$$SO_2 + \tfrac{1}{2}O_2 \rightleftharpoons SO_3$$

Its new equilibrium constant becomes:

$$K_2 = \frac{[SO_3]}{[SO_2][O_2]^{1/2}}$$

Comparing the expressions for K_1 and K_2 shows that:

$$K_2 = K_1^{1/2}$$

Therefore in general, if a reaction is multiplied by a certain factor, its equilibrium constant must be raised to a power equal to that factor in order to obtain the equilibrium constant of the new reaction.

5. If the equilibrium constant of the forward reaction of a reversible reaction is K_1, then the equilibrium constant K_2, of the backward reaction of the same reversible reaction is given by:

$$K_2 = \frac{1}{K_1}$$

6. Consider the following reactions and their equilibrium constants.

$$2NO + O_2 \rightleftharpoons 2NO_2 \qquad K_1 = \frac{[NO_2]^2}{[NO]^2[O_2]}$$

And, $\quad 2NO_2 \rightleftharpoons N_2O_4 \qquad K_2 = \dfrac{[N_2O_4]}{[NO_2]^2}$

Adding the two reactions above gives:

$$2NO + O_2 \rightleftharpoons N_2O_4 \quad K_3 = \frac{[N_2O_4]}{[NO]^2[O_2]}$$

Comparing these three equilibrium constants shows that:

$$K_3 = K_2 \times K_1$$

The relationship between K_C and K_P.

The equilibrium constant, K_P, in terms of partial pressure, and the equilibrium constant, K_C, in terms of concentration, are related by:

$$K_P = K_C(RT)^{\Delta ng}$$

where Δng is the number of moles of a gas when going from reactants to products. Δng is given by:

Δng = total number of moles of gaseous products - total number of mole of gaseous reactants

Reaction Quotient, Q_C

The reaction quotient helps us to determine if a reaction has reached equilibrium. It also tells us the direction which a reaction is progressing. The reaction quotient is calculated in the same way as the equilibrium constant.
If the reaction quotient, Q_C = equilibrium constant, K_C, then the reaction is at equilibrium. If $Q_C < K_C$, then the reaction is progressing to the right (forward reaction). If $Q_C > K_C$, then the reaction is progressing to the left (backward reaction).

Examples

1. A reversible reaction is represented as follows:

$$I_2 + I^- \rightleftharpoons I_3^-$$

The concentrations of I_2 and I^- are each equal to 0.02M at the start of reaction. If after reaction the equilibrium concentration of I_2 is 0.008M, what is the equilibrium constant of the reaction?

Solution

We are going to solve this question by establishing an ICE table. ICE means:

I: **Initial** concentration
C: **Change** in concentration
E: **Equilibrium** concentration

In order to create this table, we have to know the change in concentration of each substance in terms of x. From the reaction, the mole ratio of reactants and products is 1 : 1 : 1. Hence the change in concentration are -x, -x, +x. The substance in the direction of the reaction is given positive change as +x, while the substances on the opposite direction are given negative changes, i.e. -x. Therefore the ICE table is given as follows:

ICE	I_2	I^-	I_3^-
Initial concentration:	0.02	0.02	0
Change in concentration:	-x	-x	+x
Equilibrium concentration:	0.02 - x = 0.008	0.02 - x	x

Note that the initial concentration of I_3^- = 0. The equilibrium concentration of each substance is obtained by directly subtracting x from the initial concentration when x is negative (-x), or by directly adding x to the initial concentration when x is positive (+x).

We can solve for x by using the equilibrium concentration of I_2 as follows:

0.02 - x = 0.008   (Note that 0.008 is the equilibrium concentration of I_2 from the question)

Therefore x = 0.02 - 0.008

x = 0.012

Hence the equilibrium concentration of each of the substance is:

I_2 = 0.008

I^- = 0.008   (0.02 - x = 0.02 - 0.012 = 0.008)

I_3^- = 0.012   (i.e. x as shown on the ICE table above)

These values of equilibrium concentrations are now used to calculate the equilibrium constant as follows:

$$K_C = \frac{I_3^-}{[I_2][I^-]}$$

$$= \frac{0.012}{0.008 \times 0.008}$$

$$K_C = 187.5$$

Therefore the equilibrium constant is 187.5

2. The reversible reaction below shows the reduction of cobalt (II) oxide to produce the metal.

$$CoO + CO \rightleftharpoons Co + CO_2$$

If 1 mole each of CoO and CO are reacted in a 1 litre vessel, equilibrium is attained when 0.4 mole of each of the reactants remains. Calculate the equilibrium constant for the reaction.

Solution

The initial concentration of each of CoO and CO is given by:

$$\text{Concentration} = \frac{\text{number of moles}}{\text{volume}}$$

$$= \frac{1}{1}$$

$$= 1 \text{mol/Litre}$$

The equilibrium concentration of each of CoO and CO is given by:

$$\frac{0.4}{1} = 0.4 \text{mol/L}$$

From the reaction, the mole ration of reactants to products is, 1 : 1 : 1 : 1. Hence the changes in concentration are, -x, -x, +x, +x.. The positive values are assigned to the direction of the reaction. Therefore the ICE table is represented as follows.

ICE	CoO	CO	Co	CO_2
Initial concentration:	1	1	0	0
Change in concentration:	-x	-x	+x	+x
Equilibrium concentration:	1 - x = 0.4	1 - x = 0.4	x	x

Note that the direction of the reaction is assigned the positive changes in concentration i.e. +x.

From the table, 1 - x = 0.4. (0.4mol/L is the equilibrium concentration of CoO and CO as given in the question).

Therefore, 1 - x = 0.4

$$x = 1 - 0.4$$

$$x = 0.6$$

Hence by using the ICE table above as a guide, the equilibrium concentration of each of the substance is:

CoO = 0.4   (As given in the question)

CO = 0.4   (As given in the question)

Co = 0.6   (i.e. the value of x)

CO_2 = 0.6   (i.e. the value of x)

Therefore, the equilibrium constant, K_C, is given by:

$$K_C = \frac{[Co][CO_2]}{[CoO][CO]}$$

$$= \frac{0.6 \times 0.6}{0.4 \times 0.4}$$

$$= 2.25$$

3. Consider the reaction: $N_2 + 3H_2 \rightleftharpoons 2NH_3$ $K_C = 0.05$.

If the equilibrium concentration of nitrogen is 3.4M, and that of hydrogen is 2.2M, calculate the equilibrium concentration of ammonia.

Solution

The equilibrium constant is given by:

$$K_C = \frac{[NH_3]^2}{[N_2][H_2]^3}$$

Substituting known values into the equation above gives:

$$0.05 = \frac{x^2}{3.4 \times (2.2^3)}$$ (where x is the equilibrium concentration of NH_3)

$$0.05 = \frac{x^2}{3.4 \times 10.648}$$

$$x^2 = 0.05 \times 3.4 \times 10.648$$

$$x^2 = 1.81016$$

$$x = \sqrt{1.81016}$$

$$x = 1.35$$

Therefore the equilibrium concentration of NH_3 is 1.35M

4. 0.1mole of I_2 and 0.2mole of H_2 are reacted together in a 2 litre vessel to produce HI. If K_C for the reaction is 50, calculate the equilibrium concentration of I_2, H_2 and HI.

Solution

The equation for the reaction is given by:

$$H_2 + I_2 \rightleftharpoons 2HI$$

The initial concentration of $I_2 = \frac{0.1}{2} = 0.05$ mol/L

The initial concentration of $H_2 = \frac{0.2}{2}$

$$= 0.1 \text{mol/L}$$

The initial concentration of HI = 0 (At start of reaction, there is no product)

From the reaction, mole ratio of reactants and product is 1 : 1 : 2. Hence the changes in concentration are -x, -x, +2x. The direction of the reaction is given a positive value. The change in concentration of 2HI is +2x (positive value for direction of reaction). It is 2x since there are 2 moles of HI according to the balanced reaction. These values are now represented using an ICE table as shown below.

ICE	H_2	I_2	$2HI$
Initial concentration:	0.1	0.05	0
Change in concentration:	-x	-x	+2x
Equilibrium concentration:	0.1 - x	0.05 - x	2x

Therefore, the equilibrium constant is given by:

$$K_C = \frac{[HI]^2}{[H_2][I_2]}$$

$$50 = \frac{(2x)^2}{(0.1-x)(0.05-x)}$$

$$50 = \frac{4x^2}{0.005 - 0.1x - 0.05x + x^2}$$

$4x^2 = 50(0.005 - 0.1x - 05x + x^2)$

$4x^2 = 0.25 - 5x - 2.5x + 50x^2$

$0 = 50x^2 - 4x^2 - 7.5x + 0.25$

$46x^2 - 7.5x + 0.25 = 0$ (Quadratic equation)

Using quadratic equation formula to solve this equation gives:

$$x = \frac{-(-7.5) \pm \sqrt{(-7.5)^2 - (4 \times 46 \times 0.25)}}{2 \times 46}$$

$$= \frac{7.5 \pm \sqrt{56.25 - 46}}{92}$$

$$= \frac{7.5 \pm \sqrt{10.25}}{92}$$

$$= \frac{7.5 \pm 3.2}{92}$$

$$x = \frac{7.5 + 3.2}{92} \quad \text{Or} \quad x = \frac{7.5 - 3.2}{92}$$

x = 0.12   or   x = 0.047

x cannot be 0.12 since it cannot be greater than any of the initial concentrations.

Therefore x = 0.047

Hence the equilibrium concentrations are

H_2 = 0.1 - x

  = 0.1 - 0.047

  = 0.053 mol/L

Equilibrium concentration of I_2 = 0.05 - x

  = 0.003 mol/L

Equilibrium concentration of HI = 2x

  = 2 x 0.047

  = 0.094 mol/L

5. Consider the following reaction:

$$2H_{2(g)} + O_{2(g)} \rightleftharpoons 2H_2O_{(l)}$$

If the partial pressure of H_2 and O_2 are 0.95 atm and 1.09 atm respective, determine the equilibrium constant K_P for the reaction.

Solution

The product $H_2O_{(l)}$, is a liquid and not a gas. Therefore it is not included in the equation for the equilibrium constant. Hence, the equilibrium constant K_P is given by:

$$K_P = \frac{1}{(P_{H_2})^2 (P_{O_2})}$$ (The product part not included has been taken to be 1)

$$= \frac{1}{(0.95)^2 (1.09)}$$

$$= \frac{1}{0.983725}$$

K_P = 1.02

6. A reaction is represented by: $2N_2O_{5(g)} \rightleftharpoons O_{2(g)} + 4NO_{2(g)}$.

The mole fraction of each substance is given by: N_2O_5: 0.25, O_2: 0.32, and NO_2: 0.43. If equilibrium is established with a total pressure of 2atm, what is the equilibrium constant K_P of the reaction?

Solution

From Raoult's law, the partial pressure of each substance is given by: p = yP, where y is mole fraction, while P is total pressure.

Therefore partial pressure of N_2O_5 = 0.25 x 2   (Total pressure is 2atm)

$$= 0.5 atm$$

Partial pressure of O_2 = 0.32 x 2

$$= 0.64 atm$$

Partial pressure of NO_2 = 0.43 x 2

$$= 0.86 atm$$

Therefore the equilibrium constant is given by:

$$K_P = \frac{(P_{O_2})(P_{NO_2})^4}{(P_{N_2O_5})^2}$$

$$= \frac{(0.64)(0.86)^4}{0.5^2}$$

$$= \frac{0.64 \times 0.547}{0.25}$$

$K_P = 1.40$

7. If the total pressure of the reaction system shown below is 6atm, calculate K_P for the reaction.

$$2Cl_2O_{5(g)} \rightleftharpoons 2Cl_{2(g)} + 5O_{2(g)}$$

Solution

Molar mass of $2Cl_2O_5$ = 2[(35.5 x 2) + (5 x 16)]

$$= 2(71 + 80)$$

$$= 2 \times 151$$
$$= 302 \text{g/mol}$$

Molar mass of $2Cl_2 = 2(35.5 \times 2)$
$$= 2 \times 71$$
$$= 142 \text{g/mol}$$

Molar mass of $5O_2 = 5(16 \times 2)$
$$= 5 \times 32$$
$$= 160 \text{g/mol}$$

Since these molar masses are in g/mol, they can be used to calculate mole fraction as follows:

Total mass in g/mol = 302 + 142 + 160
$$= 604$$

Therefore, mole fraction of $Cl_2O_5 = \dfrac{302}{604}$
$$= 0.5$$

Mole fraction of $Cl_2 = \dfrac{142}{604}$
$$= 0.235$$

Mole fraction of $O_2 = \dfrac{160}{604}$
$$= 0.265$$

We now use each of the mole fractions to calculate partial pressure as follows:

Partial pressure of $Cl_2O_5 = 0.5 \times 6$ (Total pressure is 6atm)
$$= 3 \text{atm}$$

Partial pressure of $Cl_2 = 0.235 \times 6$
$$= 1.41 \text{atm}$$

Partial pressure of O_2 = 0.265 x 6

$$= 1.59 \text{atm}$$

Therefore the equilibrium constant is given by:

$$K_P = \frac{(P_{Cl_2})^2 (P_{O_2})^5}{(P_{Cl_2O_5})^2}$$

$$= \frac{(1.41)^2 (1.59)^5}{3^2}$$

$$= \frac{1.9881 \times 10.162}{9}$$

K_P = 2.24

8. Consider the reaction: $\frac{1}{2}H_{2(g)} + \frac{1}{2}I_{2(g)} \rightleftharpoons HI_{(g)}$ ΔG° = 1.7kJ/mol at 25°C.

Calculate the equilibrium constant K for the reaction at 25°C. (R = 8.314J/mol k)

Solution

ΔG° = 1.7 x 1000 = 1700J/mol, T = 25 + 273 = 298k

Therefore the equilibrium constant K is given by:

$$K_{eq} = e^{-\Delta G/RT}$$

$$= e^{-1700/(8.314 \times 298)}$$

$$= e^{-0.686}$$

K_{eq} = 0.503

9. A reaction is represented as $\frac{3}{2}H_{2(g)} + \frac{1}{2}N_{2(g)} \rightleftharpoons NH_{3(g)}$

ΔH = -45.9KJ/mol, ΔS = -99.2J/mol k at 47°C.

Determine the equilibrium constant of the reaction.

Solution

T = 47 + 273 = 320k

ΔH = -45kJ/mol

ΔS = ($\frac{-99.2}{1000}$)kJ/mol k   (Note that ΔH and ΔS should be expressed in kJ or J)

ΔS = -0.0992kJ/mol k

Recall that: ΔG = ΔH - TΔS

$\qquad$ = -45.9 - [320 x (-0.0992)]

$\qquad$ = -45.9 + 31.744

$\qquad$ ΔG = -14.156kJ/mol

$\qquad$ = -14156J/mol   (multiply kJ by 1000 to convert it to J)

Recall that: $K_{eq} = e^{-\Delta G/RT}$

$\qquad$ = $e^{-(-14156)/(8.314 \times 320)}$

$\qquad$ = $e^{14156/2660.48}$

$\qquad$ = $e^{5.321}$

K_{eq} = 204.6

Therefore the equilibrium constant is 204.6

10. The equilibrium constant for the reaction, $2N_2O_{5(g)} \rightleftharpoons O_{2(g)} + 4NO_{2(g)}$ is 3.2. What is the equilibrium constant for the reaction $N_2O_{5(g)} \rightleftharpoons ½O_{2(g)} + 2NO_{2(g)}$

Solution

$\qquad$ $2N_2O_{5(g)} \rightleftharpoons O_{2(g)} + 4NO_{2(g)}$ K_1 = 3.2 ....Reaction 1

$\qquad$ $N_2O_{5(g)} \rightleftharpoons ½O_{2(g)} + 2NO_{2(g)}$ K_2 = ? ....Reaction 2

Comparing the two reactions above, shows that Reaction 1, has been multiplied by a factor of ½ to obtain reaction 2. Hence the new equilibrium constant will be obtained by raising the old equilibrium constant to a power which is equal to the factor ½.

Therefore $K_2 = K_1^{½}$

$$= 3.2^{½}$$

$$K_2 = 1.79$$

11. Consider the following reaction: $2H_{2(g)} + O_{2(g)} \rightleftharpoons 2H_2O_{(g)}$

If the equilibrium constant for the forward reaction is 2.5, what is the equilibrium constant for the backward reaction?

Solution

Let the equilibrium constant for the forward reaction be K_1, and for the backward reaction be K_2. K_1 and K_2 are related as follows:

$$K_2 = \frac{1}{K_1}$$

$$= \frac{1}{2.5}$$

$$= 0.4$$

Therefore the equilibrium constant of the backward reaction is 0.4.

12. Two reactions and their equilibrium constants are given as follows:

$$2NO + O_2 \rightleftharpoons 2NO_2 \quad K_1 = 1.85$$

$$2NO_2 \rightleftharpoons N_2O_4 \quad K_2 = 0.92$$

From the information provided above, determine the equilibrium constant of the reaction below.

$$2NO + O_2 \rightleftharpoons N_2O_4$$

Solution

A careful observation of the reactions above shows that the third reaction was obtained by a combination of the first two reactions. Therefore, the equilibrium constant, K_3, of the third reaction can be obtained as follows:

$K_3 = K_1 \times K_2$

$= 1.85 \times 0.92$

$K_3 = 1.702$

13. An equation is represented as follows: $2SO_{2(g)} + O_{2(g)} \rightleftharpoons 2SO_{3(g)}$

If K_C for this equilibrium reaction at 30°C is 8×10^{24}, determine the value of K_P for the reaction at this temperature. (R = 8.314dm³Pa/mol k)

Solution

$T = 30 + 273$

$= 303k$

Total gaseous product moles = 2   (From $2SO_3$)

Total gaseous reactants moles = 2 + 1   (From $2SO_2$ and O_2)

$= 3$

Therefore, Δn_g = Total gaseous product moles - Total gaseous reactants moles

$= 2 - 3$

$\Delta n_g = -1$

Hence, $K_P = K_C(RT)^{\Delta n_g}$

$= 8 \times 10^{24} \times (8.314 \times 303)^{-1}$

$= \dfrac{8 \times 10^{24}}{(8.314 \times 303)^1}$

$= \dfrac{8 \times 10^{24}}{2519.142}$

$K_P = 3.18 \times 10^{21}$

14. Calculate the equilibrium constant, K_C, of the reaction below.

$CaCO_{3(s)} \rightleftharpoons CaO_{(s)} + CO_{2(g)}$ $K_P = 1.9 \times 10^{-2}$ at 500k

(R = 8.314dm³ Pa/mol k)

Solution

T = 500k

Δng = total gaseous product moles - total gaseous reactant moles

= 1 - 0   (There is no gaseous reactant, and there is only CO_2 as gaseous product)

Δng = 1

Hence, $K_P = K_C(RT)^{\Delta ng}$

Therefore, $K_C = \dfrac{K_p}{(RT)^{\Delta ng}}$

$= \dfrac{1.9 \times 10^{-2}}{(8.314 \times 500)^1}$

$K_C = 4.57 \times 10^{-6}$

15. The equilibrium constant for the reaction below is 49.2.

$$H_{2(g)} + I_{2(g)} \rightleftharpoons 2HI_{(g)}$$

If 1.5moles of H_2 and 1.5moles of I_2 are placed in a 10dm³ vessel and allowed to react, calculate the concentration of each substance at equilibrium.

Solution

Initial concentration of $H_2 = \dfrac{1.5}{10} = 0.15$mol/dm³

Initial concentration of $I_2 = \dfrac{1.5}{10} = 0.15$mol/dm³

From the reaction, the mole ratio of reactants to product is 1 : 1 : 2. Hence the changes in concentration are, -x, -x, +2x. We now use these values to present an ICE table as follows.

ICE	H_2	I_2	2HI
Initial concentration:	0.15	0.15	0
Change in concentration:	-x	-x	+2x
Equilibrium concentration:	0.15 - x	0.15 - x	2x

Therefore the equilibrium constant is given by:

$$K_c = \frac{[HI]^2}{[H_2][I_2]}$$

Substitute the respective equilibrium concentrations into the equation above. This gives:

$$K_c = \frac{(2x)^2}{(0.15 - x)(0.15 - x)}$$

$$49.2 = \frac{(2x)^2}{(0.15 - x)^2}$$

Taking the square root of both sides of the equation gives:

$$7.01 = \frac{2x}{0.15 - x}$$

$$2x = 7.01(0.15 - x)$$

$$2x = 1.05 - 7.01x$$

$$2x + 7.01x = 1.05$$

$$9.01x = 1.05$$

$$x = \frac{1.05}{9.01}$$

$$x = 0.12$$

Therefore the equilibrium concentration of each substance is given by using the ICE table above as a guide as follows.

H_2 = 0.15 - x

= 0.15 - 0.12

= 0.03 mol/dm^3

I_2 = 0.03 mol/dm^3 (Same as H_2 i.e. 0.15 - x)

HI = 2x

= 2 x 0.12

= 0.24 mol/dm^3

16. A mixture of carbon (II) oxide and steam in the proportion of 2 : 3 by volume is heated to 420°C. If at this temperature the equilibrium constant of the reaction is 1.64, determine the composition of the equilibrium mixture.

Solution

The equation for the reaction is given by: $CO_{(g)} + H_2O_{(g)} \rightleftharpoons CO_{2(g)} + H_{2(g)}$

The volumes given in the question can be used as number of reacting moles since the volume of a gas is proportional to its number of moles. The volume will now serve as our initial concentration. From the reaction, mole ratio is given by, 1 : 1 : 1 : 1. Hence the changes in concentration are, -x, -x, +x, +x. The direction of reaction is given positive values. We now use these values to create an ICE table as follows.

ICE	CO	H$_2$O	CO$_2$	H$_2$
Initial concentration:	2	3	0	0
Change in concentration:	-x	-x	+x	+x
Equilibrium concentration:	2 - x	3 - x	x	x

The equilibrium constant is given by:

$$K_C = \frac{[CO_2][H_2]}{[CO][H_2O]}$$

Substituting the respective equilibrium concentrations from the ICE table into the equilibrium constant equation above gives:

$$K_C = \frac{(x)(x)}{(2-x)(3-x)}$$

$$1.64 = \frac{x^2}{6 - 2x - 3x + x^2}$$

$1.64(6 - 5x + x^2) = x^2$

$9.84 - 8.2x + 1.64x^2 - x^2 = 0$

$0.64x^2 - 8.2x + 9.84 = 0$ (Quadratic equation)

$$x = \frac{-(-8.2) \pm \sqrt{(-8.2)^2 - (4 \times 0.64 \times 9.84)}}{2 \times 0.64}$$

$$= \frac{8.2 \pm \sqrt{67.24 - 25.1904}}{1.28}$$

$$= \frac{8.2 \pm \sqrt{42.0496}}{1.28}$$

$$= \frac{8.2 \pm 6.48}{1.28}$$

$$x = \frac{8.2 + 6.48}{1.28} \quad \text{or} \quad x = \frac{8.2 - 6.48}{1.28}$$

x = 11.47 or x = 1.34

x cannot be 11.47 since it cannot be greater than the initial concentrations of the reactants.

Therefore, x = 1.34

Hence the compositions of the equilibrium mixture in parts by volume are given by:

CO: 2 - x = 2 - 1.34

= 0.66

H_2O: 3 - x = 3 - 1.34

= 1.66

CO_2: x = 1.34

H_2: x = 1.34

17. Consider the following reaction: $CO_{2(g)} + H_{2(g)} \rightleftharpoons CO_{(g)} + H_2O_{(g)}$ $K_C = 0.137$

At a point during the reaction, the concentration of each substance was found to be: CO_2 = 5M, H_2 = 5M, CO = 1M and H_2O = 1M.

(a). At this point, in which direction is the reaction progressing?

(b). Determine the concentration of each substance at equilibrium

Solution

(a). Let us first calculate the reaction quotient, Q_C. The reaction quotient is calculated in the same way as the equilibrium constant. The reaction quotient helps us to determine if a reaction has reached equilibrium. It also tells us the direction which a reaction is progressing.

If the reaction quotient, Q_C = equilibrium constant, K_C, then the reaction is at equilibrium. If $Q_C < K_C$ (i.e. $K_C > Q_C$), then the reaction is progressing to the right (forward reaction). If $Q_C > K_C$ (i.e. $K_C < Q_C$), then the reaction is progressing to the left (backward reaction).

Therefore, $Q_C = \dfrac{[CO][H_2O]}{[CO_2][H_2]}$

$= \dfrac{1 \times 1}{5 \times 5}$

$= \dfrac{1}{25}$

$Q_C = 0.04$

From the question, $K_C = 0.137$. This shows that $Q_C < K_C$, or $K_C > Q_C$. Therefore, the reaction is progressing to the right. This means that the reaction is in the forward direction.

(b). Let us establish an ICE table for the reaction in order to calculate the equilibrium concentrations of each substance. The ICE table is as shown below.

ICE	CO_2	H_2	CO	H_2O
Initial concentration:	5	5	1	1
Change in concentration:	-x	-x	+x	+x
Equilibrium concentration:	5 - x	5 - x	1 + x	1 + x

The equilibrium constant is given by:

$K_C = \dfrac{[CO][H_2O]}{[CO_2][H_2]}$

$0.137 = \dfrac{(1+x)(1+x)}{(5-x)(5-x)}$

$0.137 = \dfrac{1 + x + x + x^2}{25 - 5x - 5x + x^2}$

$0.137 = \dfrac{1 + 2x + x^2}{25 - 10x + x^2}$

$1 + 2x + x^2 = 0.137(25 - 10x + x^2)$

$1 + 2x + x^2 = 3.425 - 1.37x + 0.137x^2$

$x^2 - 0.137x^2 + 2x + 1.37x + 1 - 3.425 = 0$

$0.863x^2 + 3.37x - 2.425 = 0$

Using quadratic equation formula, we obtain x as follows:

$$x = \frac{-3.37 \pm \sqrt{(-3.37)^2 - [4 \times 0.863 \times (-2.425)]}}{2 \times 0.863}$$

$$= \frac{-3.37 \pm \sqrt{11.357 + 8.371}}{1.726}$$

$$= \frac{-3.37 \pm \sqrt{19.728}}{1.726}$$

$$= \frac{-3.37 \pm 4.442}{1.726}$$

$$= \frac{-3.37 + 4.442}{1.726}$$

$$= \frac{1.072}{1.726} \quad \text{(The negative value has been neglected since x cannot be negative)}$$

x = 0.621.

Therefore, the equilibrium concentrations are:

CO_2: 5 - x = 5 - 0.621

= 4.379M

H_2: 5 - x = 5 - 0.621

= 4.379M

CO: 1 + x = 1 + 0.621

= 1.621

H_2O: 1 + x = 1 + 0.621

= 1.621

18. Consider the reaction below:

$$NH_4CO_2NH_{2(s)} \rightleftharpoons CO_{2(g)} + 2NH_{3(g)}$$

At room temperature, the total pressure of the gases in equilibrium with the solid is 0.14atm. Calculate the equilibrium constant of the reaction.

Solution

Let us calculate the molar masses of the gaseous products. The solid reactant is not included in the calculation of the equilibrium constant since pressure in involved.

Therefore, $2NH_3 = 2[14 + (1 \times 3)]$

$$= 2(17)$$

$$= 34g/mol$$

$CO_2 = 12 + (16 \times 2)$

$$= 44g/mol$$

These masses in g/mol can be used to calculate the mole fraction of each of the gas as follows.

Total mass in g/mol = 34 + 44

$$= 78g/mol$$

Therefore, mole fraction of $NH_3 = \dfrac{34}{78}$

$$= 0.436$$

Mole fraction of $CO_2 = \dfrac{44}{78}$

$$= 0.564$$

Therefore by using Raoult's law (p = yP), the partial pressure of each product can be determined as follows:

Partial pressure of NH_3 = 0.436 x 0.14   (Total pressure = 0.14atm)

$$= 0.061atm$$

Partial pressure of CO_2 = 0.564 x 0.14

$$= 0.079atm$$

Therefore the equilibrium constant is given by:

$$K_P = \frac{(P_{CO_2})(P_{NH_3})^2}{1}$$ (The solid is not included, hence the 1 as our denominator)

$$= \frac{(0.079)(0.061)^2}{1}$$

$$= 0.079 \times 0.003721$$

$$K_P = 2.94 \times 10^{-4}$$

19. Consider the following reaction: $H_{2(g)} + CO_{2(g)} \rightleftharpoons CO_{(g)} + H_2O_{(g)}$ $K_C = 0.2$ at 720k.

If 0.5mole of H_2 and 0.5mole of CO_2 are mixed in a 5 litres vessel at 720k, determine the equilibrium pressure of each substance.

Solution

Initial concentration of $H_2 = \dfrac{0.5}{5}$

$$= 0.1 \text{mol/litre}$$

Initial concentration of $CO_2 = \dfrac{0.5}{5}$

$$= 0.1 \text{mol/litre}$$

The changes in concentration of reactants and products are, -x, -x, +x and +x, respectively according to the mole ratio of the reaction. Hence the ICE table is given as follows.

ICE	H_2	CO_2	CO	H_2O
Initial concentration:	0.1	0.1	0	0
Change in concentration:	-x	-x	+x	+x
Equilibrium concentration:	0.1 - x	0.1 - x	x	x

Hence the equilibrium constant is given by:

$$K_C = \frac{(x)(x)}{(0.1-x)(0.1-x)}$$

$$0.2 = \frac{x^2}{(0.1-x)^2}$$

Taking the square root of both sides of the equation gives:

$$0.447 = \frac{x}{(0.1-x)}$$

$$x = 0.447(0.1 - x)$$

$$x = 0.0447 - 0.447x$$

$$x + 0.447x = 0.0447$$

$$1.447x = 0.0447$$

$$x = \frac{0.0447}{1.447}$$

$$x = 0.0309$$

Therefore the equilibrium concentration of each substance is given by:

H_2: 0.1 - x = 0.1 - 0.0309

= 0.0691mol/L

CO_2: 0.1 - 0.0309 = 0.0691mol/L

CO; x = 0.0309

H_2O: x = 0.0309

Recall that: p = mRT, where p = pressure, m = molar concentration, R = 0.0821Latm/mol k, and T = 720.

Therefore pressure of H_2 is: p = mRT

= 0.0691 x 0.0821 x 720

= 4.08atm

Pressure of CO_2 = 4.08atm   (The same as that of H_2)

Pressure of CO is: p = mRT

= 0.0309 x 0.0821 x 720

= 1.83atm

Pressure of H₂O: 1.83atm   (Same as CO)

20. A 1 litre flask contains the mixture below in equilibrium.

$$CO + Cl_2 \rightleftharpoons COCl_2$$

CO = 0.05mole, Cl_2 = 0.2mole, $COCl_2$ = 0.2mole.

(a). Calculate the equilibrium constant of the reaction

(b). If 0.1mole of CO is added, what will be the new concentrations of each component when equilibrium is re-established.

<u>Solution</u>

(a) The equilibrium concentration of each substance is given as follows:

$$CO: \frac{0.05}{1} = 0.05 mol/L$$

$$Cl_2: \frac{0.2}{1} = 0.2 mol/L$$

$$COCl_2: \frac{0.2}{1} = 0.2 mol/L$$

Therefore the equilibrium constant is calculated from these values as given below.

$$K_c = \frac{[COCl_2]}{[CO][Cl_2]}$$

$$= \frac{0.2}{0.05 \times 0.2}$$

$$= \frac{0.2}{0.01}$$

$$K_c = 20$$

(b). When 0.1mole of CO is added to the system, according to Le Chatelier's principle, the system will shift so as to annul this effect. This means that the equilibrium constant will not change. In this case, the previous equilibrium concentration of each substance becomes the initial concentrations. However, the initial concentration of CO will increase to become: 0.1 + 0.05 = 0.15mol/L. The reaction will take place in the forward direction since it was a reactant that was added. Hence from the mole ratio of the reaction components, the changes in concentration will be, -x, -x and +x respectively. We can now present the ICE table as follows:

ICE	CO	Cl_2	$COCl_2$
Initial concentration:	0.15	0.2	0.2
Change in concentration:	-x	-x	+x
Equilibrium concentration:	0.15 - x	0.2 - x	0.2 + x

Hence the equilibrium constant is given by:

$$K_c = \frac{(0.2 + x)}{(0.15 - x)(0.2 - x)}$$

$$20 = \frac{(0.2 + x)}{0.03 - 0.15x - 0.2x + x^2}$$

$20(0.03 - 0.35x + x^2) = 0.2 + x$

$0.6 - 7x + 20x^2 = 0.2 + x$

$20x^2 - 7x - x + 0.6 - 0.2 = 0$

$20x^2 - 8x + 0.4 = 0$ (Quadratic equation)

Solving this quadratic equation gives:

$$x = \frac{-(-8) \pm \sqrt{(-8)^2 - (4 \times 20 \times 0.4)}}{2 \times 20}$$

$$= \frac{8 \pm \sqrt{64 - 32}}{40}$$

$$= \frac{8 \pm \sqrt{32}}{40}$$

$$= \frac{8 \pm 5.66}{40}$$

$$x = \frac{8 + 5.66}{40} \quad \text{or} \quad x = \frac{8 - 5.66}{40}$$

x = 0.342   or   x = 0.0585

But, x cannot be greater than any of the initial concentrations. Therefore, 0.342 cannot be the answer.

Hence, x = 0.0585

Therefore the new concentrations of each component when equilibrium is established again are as given below:

CO: 0.15 - x = 0.15 - 0.0585

$\qquad$ = 0.0915mol/L

Cl_2: 0.2 - x = 0.2 - 0.0585

$\qquad$ = 0.142mol/L

$COCl_2$: 0.2 + x = 0.2 + 0.0585

$\qquad$ = 0.259mol/L

EXERCISE

1. A reversible reaction is represented as follows:

$$I_2 + I^- \rightleftharpoons I_3^-$$

The concentrations of I_2 and I^- are each equal to 0.1M at the start of reaction. If after reaction the equilibrium concentration of I_2 is 0.02M, what is the equilibrium constant of the reaction?

2. The reversible reaction below shows the reduction of copper (II) oxide to produce the metal.

$$CuO + CO \rightleftharpoons Cu + CO_2$$

If 2 mole each of CuO and CO are reacted in a 10 litre vessel, equilibrium is attained when 0.9 mole of each of the reactants remains. Calculate the equilibrium constant for the reaction.

3. Consider the reaction: $N_2 + 3H_2 \rightleftharpoons 2NH_3$ $\quad K_C = 0.04$.

If the equilibrium concentration of nitrogen is 1.6M, and that of hydrogen is 0.7M, calculate the equilibrium concentration of ammonia.

4. 1.2mole of I_2 and 1.8mole of H_2 are reacted together in a 5 litre vessel to produce HI. If K_C for the reaction is 52, calculate the equilibrium concentration of I_2, H_2 and HI.

5. Consider the following reaction:

$$2H_{2(g)} + O_{2(g)} \rightleftharpoons 2H_2O_{(l)}$$

If the partial pressure of H_2 and O_2 are 0.5atm and 1.1atm respective, determine the equilibrium constant K_P for the reaction.

6. A reaction is represented by: $2N_2O_{5(g)} \rightleftharpoons O_{2(g)} + 4NO_{2(g)}$.

The mole fraction of each substance is given by: N_2O_5: 0.45, O_2: 0.12, and NO_2: 0.43. If equilibrium is established with a total pressure of 1.6atm, what is the equilibrium constant K_P of the reaction?

7. If the total pressure of the reaction system shown below is 5atm, calculate K_P for the reaction.

$$2Cl_2O_{5(g)} \rightleftharpoons 2Cl_{2(g)} + 5O_{2(g)}$$

(Cl = 35.5, O = 16)

8. Consider the reaction: $½H_{2(g)} + ½I_{2(g)} \rightleftharpoons HI_{(g)}$ $\Delta G° = 2.1kJ/mol$ at 60°C.

Calculate the equilibrium constant K for the reaction at 60°C. (R = 8.314J/mol k)

9. A reaction is represented as $\frac{3}{2}H_{2(g)} + ½N_{2(g)} \rightleftharpoons NH_{3(g)}$

$\Delta H = -44.3KJ/mol$, $\Delta S = -102J/mol\ k$ at 85°C.

Determine the equilibrium constant of the reaction.

10. The equilibrium constant for the reaction, $2N_2O_{5(g)} \rightleftharpoons O_{2(g)} + 4NO_{2(g)}$ is 0.96. What is the equilibrium constant for the reaction $N_2O_{5(g)} \rightleftharpoons ½O_{2(g)} + 2NO_{2(g)}$
(Take R as 8.314 J/mol k)

11. Consider the following reaction: $2H_{2(g)} + O_{2(g)} \rightleftharpoons 2H_2O_{(g)}$

If the equilibrium constant for the forward reaction is 1.42, what is the equilibrium constant for the backward reaction?

12. Two reactions and their equilibrium constants are given as follows:

$2NO + O_2 \rightleftharpoons 2NO_2 \quad K_1 = 1.27$

$2NO_2 \rightleftharpoons N_2O_4 \quad K_2 = 1.65$

From the information provided above, determine the equilibrium constant of the reaction below.

$2NO + O_2 \rightleftharpoons N_2O_4$

13. An equation is represented as follows: $2SO_{2(g)} + O_{2(g)} \rightleftharpoons 2SO_{3(g)}$

If K_C for this equilibrium reaction at 25°C is 5×10^{22}, determine the value of K_P for the reaction at this temperature. (R = 8.314dm³Pa/mol k)

14. Calculate the equilibrium constant, K_C, of the reaction below.

$CaCO_{3(s)} \rightleftharpoons CaO_{(s)} + CO_{2(g)} \quad K_P = 2 \times 10^{-4}$ at 410k

(R = 8.314dm³ Pa/mol k)

15. The equilibrium constant for the reaction below is 50.4.

$H_{2(g)} + I_{2(g)} \rightleftharpoons 2HI_{(g)}$

If 2.2moles of H_2 and 2.2moles of I_2 are placed in a 2dm³ vessel and allowed to react, calculate the concentration of each substance at equilibrium.

16. A mixture of carbon (II) oxide and steam in the proportion of 1 : 3 by volume is heated to 600°C. If at this temperature the equilibrium constant of the reaction is 3.8, determine the composition of the equilibrium mixture.

17. Consider the following reaction: $CO_{2(g)} + H_{2(g)} \rightleftharpoons CO_{(g)} + H_2O_{(g)} \quad K_C = 0.096$

At a point during the reaction, the concentration of each substance was found to be: CO_2 = 2M, H_2 = 2M, CO = 0.5M and H_2O = 0.5M.

(a). At this point, in which direction is the reaction progressing?

(b). Determine the concentration of each substance at equilibrium

18. Consider the reaction below:

$NH_4CO_2NH_{2(s)} \rightleftharpoons CO_{2(g)} + 2NH_{3(g)}$

At room temperature, the total pressure of the gases in equilibrium with the solid is 0.214atm. Calculate the equilibrium constant of the reaction.

19. Consider the following reaction: $H_{2(g)} + CO_{2(g)} \rightleftharpoons CO_{(g)} + H_2O_{(g)}$ $K_C = 1.5$ at 540k.

If 1.8mole of H_2 and 1.5mole of CO_2 are mixed in a 5 litres vessel at 540k, determine the equilibrium pressure of each substance.
(R = 0.0821 L atm/mol k)

20. A 5 litre flask contains the mixture below in equilibrium.

$$CO + Cl_2 \rightleftharpoons COCl_2$$

CO = 2moles, Cl_2 = 3moles, $COCl_2$ = 4moles.

(a). Calculate the equilibrium constant of the reaction

(b). If 2.5moles of Cl_2 is added to the flask, what will be the new concentrations of each component when equilibrium is re-established.

ANSWERS TO THE EXERCISES

Exercise 1
1(a) X = 0.230kmols, Y = 0.0806kmols, Z = 0.105kmols   (b) X = 0.553, Y = 0.194, Z = 0.253
(c) X = 0.445, Y = 0.210, Z = 0.345   2(a) CH_4 = 0.422, C_4H_{10} = 0.409, C_3H_8 = 0.169
(b) CH_4 = 675.2g, C_4H_{10} = 2372.2g, C_3H_8 = 743.6g
(c) CH_4 = 0.178, C_4H_{10} = 0.626, C_3H_8 = 0.196
3. CO_2 = 0.240, CO = 0.0954, O_2 = 0.233, N_2 = 0.432   4(a) NO_2 = 40.25g, NO = 51.75g
(b) NO_2 = 0.337, NO = 0.663
5. X = 9.3g/mol, Y = 81.5g/mol, Z = 17.1g/mol   6. 0.281

Exercise 2
1. 32.5g/mol   2. 7.37g/mol   3. 188g/mol   4. Ar = 0.542, Cl_2 = 0.458
5. 28.83g/mol

Exercise 3
There is no exercise to chapter 3

Exercise 4
1. 1067.7Kg   2(a) 0.691Kg   (b) Fabric: 90.6%, Moisture: 9.4%   3. 350,541.2Kg/h
4(a) 3.61Kg   (b) 1.39Kg   5. 365.5Kg

Exercise 5
1. 1,428.6kg   2(a) 677.4kg   (b) 1437.1kg   3. $56.27
4. 98% HCl = 13.7litres, 64% HCl = 4.9litres   5. $8.5

Exercise 6
1. 6.3%   2. 34.4%   3. Required air = 571.4kmols, while actual air supplied = 860kmols
4(a) C = 88%, H_2 = 12%   (b) 50.4%
5. CO_2 = 8.00%, H_2O = 12.42%, CO = 0.87%, O_2 = 4.77%, N_2 = 73.96%
6(a) N_2 = 71.71%, CO_2 = 20.42%, H_2O = 7.40%, SO_2 = 0.48%   (b) 1965.3m^3
7. N_2 = 73.97%, C_3H_8 = 0.79%, O_2 = 5.72%, CO_2 = 8.37%, H_2O = 11.15%

Exercise 7
1(a) Fe   (b) 25%   (c) 89.4%   2(a) O_2 (b) 25%   (c) Ca = 20%, CaO = 80%
(d) Ca = 26.9%, O2 = 3.8%, CaO = 69.2%   3(a) C   (b) 12.9%   (c) 95.6%
4(a) Fe   (b) 101.4%   (c) 96.4%   5(a) C   (b) 86.8%   (c) 94.2%

Exercise 8
1(a) Top product = 430.4kg/h, Bottom product = 1569.6kg/h   (b) 404.6kg/h
(c) 235.4kg/h   (d) 1.9   2(a) 32.3kg   (b) 35.9%   (c) 80.8%   (d) 56.4kg

3. 1.478 x 10^{-3} m^3/min   4(a) 182.6m^3/min   (b) 0.173kmols/min   5. 2450.2m^3
6(a) 325.9kg/h   (b) 3960.9m^3/h   7(a) 1394.4kg/h   (b) 524.8kgH$_2$O/h

Exercise 9
1. 11743kg/h   2. 0.24%   3. 3294kg/h   4. 250kg   5. 453.7kg

Exercise 10
1. Cu$_2$O   2(a) O = 50%, S = 50%   (b) SO$_2$ 2. Na$_2$CO$_3$.10H$_2$O   4(a) C$_4$H$_8$
(b) 72cm^3 5. C$_5$H$_{10}$ 6. C$_3$H$_6$ 7(a) C$_2$H$_6$ (b) 58cm^3 8(a) CH
(b) C$_2$H$_2$ 9(a) C$_5$H$_{12}$ (b) C$_5$H$_{12}$ 10(a) CH$_4$ (b) CH$_4$

Exercise 11
1 (a) 437.5kg/m^3 (b) 583.3kg/m^3 2(a) 18.6m   (b) 23.8m   3(a) 2818kg/m^3
(b) 53.5mmHg   (c) 686.5mmHg   4(a) $\Delta P = (\rho_g - \rho_p)gh_1 + (\rho_r - \rho_p)gh_2$
(b) 161325.4N/m2   5. 58.1mmHg

Exercise 12
1 y_{H_2O} = 0.2818, y_{air} = 0.7182   2(a) 0.0342   (b) 0.0354   (c) 25.1°C   (d) 0.02198
3. 1.0litres   4(a) 982.8g   (b) 47.2litres
5(a) y_{H_2O} = 0.00936   (b) 0.00945   (c) 5.9°C   (d) 0.00587

Exercise 13
1. 200   2. 1.46   3. 0.148   4. I$_2$ = 0.025mol/L, H$_2$ = 0.145mol/L, HI = 0.43mol/L
5. 3.64   6. 0.0830   7. 0.902   8. 0.468   9. 18.65   10. 0.980   11. 0.704
12. 2.10   13. 2.02 x 10^{19} 14. 5.87 x 10^{-8} 15. H$_2$ = 0.24, I$_2$ = 0.24, HI = 1.72
16. CO = 0.101, H$_2$O = 2.101, CO$_2$ = 0.899, H$_2$ = 0.899   17(a) Forward reaction
(b) CO$_2$ = 1.91M, H$_2$ = 1.91M, CO = 0.592M, H$_2$O = 0.592M   18. 1.05 x 10^{-3}
19. H$_2$ = 7.98atm, CO$_2$ = 5.32atm, CO = 7.98atm, H$_2$O = 7.98atm
20(a) 3.33   (b) CO = 0.281mol/L, Cl$_2$ = 0.981mol/L, COCl$_2$ = 0.919mol/L

For other books written by the author, go to amazon and search for the authors name: Kingsley Augustine. You will see all the books written by the author.

For issues, enquiries and suggestions as regards this book, contact the author on:
Email: kingzohb2@yahoo.com

www.ingramcontent.com/pod-product-compliance
Lightning Source LLC
Chambersburg PA
CBHW082247220526
45469CB00009B/2908